Universitext

Springer

New York
Berlin
Heidelberg
Barcelona
Budapest
Hong Kong
London
Milan
Paris
Santa Clara
Singapore
Tokyo

Universitext

Aksoy/Khamsi: Nonstandard Methods in Fixed Point Theory
Aupetit: A Primer on Spectral Theory
Booss/Bleecker: Topology and Analysis
Borkar: Probability Theory: An Advanced Course
Carleson/Gamelin: Complex Dynamics
Cecil: Lie Sphere Geometry: With Applications to Submanifolds
Chae: Lebesgue Integration (2nd ed.)
Charlap: Bieberbach Groups and Flat Manifolds
Chern: Complex Manifolds Without Potential Theory
Cohn: A Classical Invitation to Algebraic Numbers and Class Fields
Curtis: Abstract Linear Algebra
Curtis: Matrix Groups
DiBenedetto: Degenerate Parabolic Equations
Dimca: Singularities and Topology of Hypersurfaces
Edwards: A Formal Background to Mathematics I a/b
Edwards: A Formal Background to Mathematics II a/b
Foulds: Graph Theory Applications
Fuhrmann: A Polynomial Approach to Linear Algebra
Gardiner: A First Course in Group Theory
Gårding/Tambour: Algebra for Computer Science
Goldblatt: Orthogonality and Spacetime Geometry
Hahn: Quadratic Algebras, Clifford Algebras, and Arithmetic Witt Groups
Holmgren: A First Course in Discrete Dynamical Systems
Howe/Tan: Non-Abelian Harmonic Analysis: Applications of $SL(2, R)$
Howes: Modern Analysis and Topology
Humi/Miller: Second Course in Ordinary Differential Equations
Hurwitz/Kritikos: Lectures on Number Theory
Jennings: Modern Geometry with Applications
Jones/Morris/Pearson: Abstract Algebra and Famous Impossibilities
Kannan/Krueger: Advanced Analysis
Kelly/Matthews: The Non-Euclidean Hyperbolic Plane
Kostrikin: Introduction to Algebra
Luecking/Rubel: Complex Analysis: A Functional Analysis Approach
MacLane/Moerdijk: Sheaves in Geometry and Logic
Marcus: Number Fields
McCarthy: Introduction to Arithmetical Functions
Meyer: Essential Mathematics for Applied Fields
Mines/Richman/Ruitenburg: A Course in Constructive Algebra
Moise: Introductory Problems Course in Analysis and Topology
Morris: Introduction to Game Theory
Porter/Woods: Extensions and Absolutes of Hausdorff Spaces
Ramsay/Richtmyer: Introduction to Hyperbolic Geometry
Reisel: Elementary Theory of Metric Spaces
Rickart: Natural Function Algebras
Rotman: Galois Theory

(continued after index)

Chuanming Zong

Strange Phenomena in Convex and Discrete Geometry

Edited by James J. Dudziak

Springer

Chuanming Zong
Institute of Mathematics
The Chinese Academy of Sciences
Beijing 100080, PR China

James J. Dudziak
Department of Mathematics
Bucknell University
Lewisburg, PA 17837
USA

With 10 illustrations.

Mathematics Subject Classification (1991): 52-01

Library of Congress-in-Publication Data
Zong, Chuanming.
Strange phenomena in convex and discrete geometry / Chuanming Zong.
p. cm. – (Universitext)
Includes bibliographical references and index.
ISBN 0-387-94734-5 (soft: alk. paper)
1. Convex geometry. 2. Combinatorial geometry. I. Title.
QA639.5.Z66 1996
516'.08–dc20 96-11737

Printed on acid-free paper.

Production managed by Frank Ganz; manufacturing supervised by Jeffrey Taub.
Camera-ready copy prepared from the author's LaTeX files.
Printed and bound by Braun-Brumfield, Inc., Ann Arbor, MI.
Printed in the United States of America.

9 8 7 6 5 4 3 2 1

ISBN 0-387-94734-5 Springer-Verlag New York Berlin Heidelberg SPIN 10524268

Preface

Convex and Discrete Geometry, in the sense used by many mathematicians, is one of the most intuitive subjects in mathematics. It has the characteristic that many of its hardest problems, such as the sphere packing problem or Borsuk's problem, can be explained, along with their conjectured answers, to a layman in a few minutes. However, proofs of the conjectured answers to some of these simply stated problems often have cost the best mathematicians decades or centuries of effort. More surprisingly, some of these commonly believed conjectures, whose truth seemed intuitively certain, were not true. The conjectured answer to Borsuk's problem is an example. Furthermore, there are problems in Convex and Discrete Geometry whose answers are so counterintuitive and strange that they can hardly be believed before reading their proofs. The purpose of this book is to present just some of the most famous problems in Convex and Discrete Geometry which possess such incredible answers.

Although this book deals with difficult problems and presents some of the most recent advances in Convex and Discrete Geometry, it is self-contained and can be understood by any trained mathematician.

For invaluable help and consultation, I am obliged to Professors M. Berger, J. Dudziak, P.M. Gruber, E. Hlawka, D. Larman, P. Mani-Levitska, R. Pollack, R. Schneider, Y. Wang, J.M. Wills, and G. Xu. Professor J. Dudziak's thorough editorial work has improved the quality of the final manuscript. Nevertheless, the responsibility is wholly mine for any mistakes and faults of exposition that still remain in the book. The staff at Springer-Verlag in New York have been courteous, competent, and helpful, especially Mr. T. von Foerster and Ms. J. Wolkowicki. This work was supported by the National Scientific Foundation of China. Its final revision was done while I was a guest at IHES, ETH Zürich, and UCL. I am very grateful to these institutions for their support and hospitality.

London 1996 *C. Zong*

Basic Notation

For the convenience of the reader, we list here the following notation, which will be used throughout this book:

R^n: Euclidean space of n-dimensions.

$\mathbb{Z}$: The set of all integers.

x or x_i: Points in R^n.

x^j: The j-th coordinate of x.

$\text{conv}\{X\}$: The convex hull of a given set X. Thus,

$$\text{conv}\{X\} = \left\{\sum \lambda_i x_i : \text{each } x_i \in X, \text{ each } \lambda_i \geq 0, \text{ and } \sum \lambda_i = 1\right\}.$$

$\langle x, y\rangle$: The inner product of x and y.

$\|x - y\|$: The Euclidean distance between x and y.

$d(X)$: The diameter of X. Thus,

$$d(X) = \sup_{x,y \in X} \|x - y\|.$$

$d(X, Y)$: The Euclidean distance between sets X and Y. Thus,

$$d(X, Y) = \min_{x \in X,\ y \in Y} \|x - y\|.$$

K: An n-dimensional convex body, i.e., a compact convex set in R^n with nonempty interior.

$\partial(K)$: The boundary of K.

$\text{int}(K)$: The interior of K.

$v(K)$: The volume of K.

$s(K)$: The surface area of K.
$D(K)$: The difference body of K. Thus, $D(K) = \{x - y: x, y \in K\}$.
$\delta^H(K_1, K_2)$: The Hausdorff distance between K_1 and K_2. Thus,

$$\delta^H(K_1, K_2) = \max\left\{ \sup_{x\in K_1} \inf_{y\in K_2} \|x - y\|,\ \sup_{y\in K_2} \inf_{x\in K_1} \|x - y\| \right\}.$$

$\mathcal{K}$: The space of all n-dimensional convex bodies with the Hausdorff metric δ^H.
B: The n-dimensional unit ball centered at the origin o.
ω_n: The volume of B. Thus,

$$\omega_n = \frac{\pi^{\frac{n}{2}}}{\Gamma(1 + \frac{n}{2})}.$$

W: The n-dimensional unit cube $\{(x^1, x^2, \ldots, x^n): \|x^i\| \le \frac{1}{2},\ 1 \le i \le n\}$.
S: An n-dimensional simplex.
P: An n-dimensional polytope.
C: An n-dimensional centrally symmetric convex body.
$\mathcal{C}$: The space of all n-dimensional centrally symmetric convex bodies with the Hausdorff metric δ^H.
Λ: An n-dimensional lattice.

Contents

1

Borsuk's Problem

§1. Introduction

Let X denote a subset of R^n. As usual, we call

$$d(X) = \sup_{x,y \in X} \|x - y\|$$

the *diameter* of X. In studying the relation between a set and its subsets of smaller diameter, K. Borsuk [2] in 1933 raised the following famous problem:

Borsuk's Problem. *Is it true that every bounded set X in R^n can be partitioned into $n+1$ subsets $X_1, X_2, \ldots, X_{n+1}$ such that*

$$d(X_i) < d(X), \quad i = 1,\ 2, \ldots, n+1?$$

An application of the *pigeonhole principle* to the case where X is a regular n-dimensional simplex shows that fewer than $n+1$ subsets of smaller diameter do not suffice to partition X. Thus, $n+1$ is the smallest number that might suffice in general.

During the last six decades, this simply stated problem has been studied by many mathematicians. Although nobody could provide a proof, it was widely believed that the answer to the problem was "yes." Surprisingly, J. Kahn and G. Kalai [1] have recently proved that for sufficiently large values of the dimension of the Euclidean space, the answer is "no." In this chapter we will first discuss the affirmative answer for $n = 3$ due to J. Perkal [1] and H.G. Eggleston [1], and then the negative answer for sufficiently large n due to J. Kahn and G. Kalai [1].

§2. The Perkal-Eggleston Theorem

Theorem 1.1 (J. Perkal [1] and H.G. Eggleston [1]). *Every bounded 3-dimensional set X can be divided into four parts, X_1, X_2, X_3, and X_4, such that*

$$d(X_i) < d(X), \quad i = 1,\ 2,\ 3,\ 4.$$

First, we prove a general lemma which is not only necessary for the proof of this theorem but also interesting in itself.

Lemma 1.1 (D. Gale [1]). *In R^n, every set X with $d(X) = 1$ can be inscribed in a regular simplex S with*

$$d(S) \leq \left(\frac{n(n+1)}{2}\right)^{\frac{1}{2}}.$$

Proof: Let $e_0, e_1, \ldots, e_n$ be a standard basis of R^{n+1}. Take $e = \sum_{i=0}^{n} e_i$ and $H = \{x \in R^{n+1} \colon \langle x, e\rangle = 0\}$. Since H is an n-dimensional linear subspace of R^{n+1}, we may assume that X is embedded in H. We may also assume, without loss of generality, that X is closed. Defining

$$\alpha_i = \min_{x \in X} \langle x, e_i\rangle, \quad \beta_i = \max_{x \in X} \langle x, e_i\rangle,$$

$$H_i' = \{x \in R^{n+1} : \ \langle x, e_i\rangle \geq \alpha_i\},$$

$$H_i^* = \{x \in R^{n+1} : \ \langle x, e_i\rangle \leq \beta_i\},$$

$$a = \sum_{i=0}^{n} \alpha_i e_i, \quad b = \sum_{i=0}^{n} \beta_i e_i,$$

$$S' = \bigcap_{i=0}^{n} H_i' \cap H, \quad \text{and} \quad S^* = \bigcap_{i=0}^{n} H_i^* \cap H,$$

we proceed to show that both S' and S^* are regular simplices which circumscribe X and that at least one of them has diameter $d \leq \left(\frac{n(n+1)}{2}\right)^{\frac{1}{2}}$. The proof is divided into three parts.

First, we show that $\langle a, e\rangle < 0$ and $\langle b, e\rangle > 0$.

It follows from the previous definitions that for any $x \in X$,

$$\langle a, e\rangle = \sum_{i=0}^{n} \alpha_i \leq \sum_{i=0}^{n} \langle x, e_i\rangle = \langle x, e\rangle = 0.$$

Hence, if $\langle a, e\rangle = 0$, then one must have, for any $x \in X$, that

$$x = \sum_{i=0}^{n} \langle x, e_i\rangle e_i = \sum_{i=0}^{n} \alpha_i e_i = a.$$

But then X is a singleton, which contradicts the hypothesis that $d(X) = 1$. We conclude that $\langle a, e\rangle < 0$ and similarly, that $\langle b, e\rangle > 0$.

Second, we prove that S' and S^* are simplices in which X is inscribed.

To do this, we show that S' is the simplex spanned by the $n+1$ vectors $v_i = a - \langle a, e\rangle e_i$, $i = 0, 1, \ldots, n$. Suppose that $x = \sum_{i=0}^{n} \lambda_i v_i$ with $\lambda_i \geq 0$ and $\sum_{i=0}^{n} \lambda_i = 1$. Then

$$x = \sum_{i=0}^{n} \lambda_i (a - \langle a, e\rangle e_i) = a - \langle a, e\rangle \sum_{i=0}^{n} \lambda_i e_i.$$

Hence, keeping $\langle a, e\rangle < 0$ in mind, we get

$$\langle x, e_i\rangle = \alpha_i - \lambda_i \langle a, e\rangle \geq \alpha_i$$

and

$$\langle x, e\rangle = \sum_{i=0}^{n} \langle x, e_i\rangle = \sum_{i=0}^{n} \alpha_i - \langle a, e\rangle = 0,$$

which imply that $x \in S'$.

On the other hand, suppose that $x \in S'$. Then we have

$$x = \sum_{i=0}^{n} \gamma_i e_i, \quad \gamma_i \geq \alpha_i, \quad \sum_{i=0}^{n} \gamma_i = 0.$$

Since $\langle a, e\rangle \neq 0$, we can write $e_i = \frac{a - v_i}{\langle a, e\rangle}$. Hence, we have

$$x = \frac{1}{\langle a, e\rangle} \sum_{i=0}^{n} \gamma_i (a - v_i) = -\frac{1}{\langle a, e\rangle} \sum_{i=0}^{n} \gamma_i v_i.$$

By the definition of v_i we see that $\sum_{i=0}^{n} \alpha_i v_i = 0$. So we may write

$$x = -\frac{1}{\langle a, e\rangle} \sum_{i=0}^{n} \gamma_i v_i + \frac{1}{\langle a, e\rangle} \sum_{i=0}^{n} \alpha_i v_i = \frac{1}{\langle a, e\rangle} \sum_{i=0}^{n} (\alpha_i - \gamma_i) v_i. \tag{1.1}$$

Taking $\zeta_i = \frac{\alpha_i - \gamma_i}{\langle a, e\rangle}$, since $\gamma_i \geq \alpha_i$ and $\langle a, e\rangle < 0$, it is easy to see that $\zeta_i \geq 0$ and $\sum_{i=0}^{n} \zeta_i = 1$. Therefore, it follows from (1.1) that x belongs to the *convex hull* of the $n+1$ points $v_0, v_1, \ldots, v_n$. Hence, $S' = \operatorname{conv}\{v_0, v_1, \ldots, v_n\}$.

To prove that X is inscribed in S' we must show that X has a point in common with each *facet* of S'. Since there is an $x \in X$ such that $\langle x, e_i\rangle =$

$\gamma_i = \alpha_i$, we see by (1.1) that this x lies in the facet of S' opposite the vertex v_i.

The proof of the corresponding assertion for S^* is similar.

Finally, we prove that S' and S^* are regular simplices with diameters $d' = -\sqrt{2}\langle a, e\rangle$ and $d^* = \sqrt{2}\langle b, e\rangle$, respectively, and that

$$\min\{d', d^*\} \leq \left(\frac{n(n+1)}{2}\right)^{\frac{1}{2}}.$$

Note that $\|v_i - v_j\| = |\langle a, e\rangle| \cdot \|e_i - e_j\| = -\sqrt{2}\langle a, e\rangle$ whenever $i \neq j$; thus, S' is regular. Since $d(\text{conv}\{X\}) = d(X)$ for any set X, we see that $d' = -\sqrt{2}\langle a, e\rangle$. Similarly, S^* is regular and $d^* = \sqrt{2}\langle b, e\rangle$.

To estimate $\min\{d', d^*\}$, we first show that $\beta_i - \alpha_i \leq \left(\frac{n}{n+1}\right)^{\frac{1}{2}}$ holds for all i. Without loss of generality, we show this only for $i = 0$.

Choose $y, z \in X$ such that $\langle y, e_0\rangle = \alpha_0$ and $\langle z, e_0\rangle = \beta_0$ and let

$$\theta_i = \langle z - y, e_i\rangle, \quad i = 0,\ 1, \ldots, n.$$

Keeping the hypothesis $d(X) = 1$ in mind, it is easy to see that

$$\sum_{i=0}^{n} \theta_i = 0 \quad \text{and} \quad \sum_{i=0}^{n} \theta_i^2 \leq 1.$$

Then, we have

$$\begin{aligned}\sum_{i=1}^{n} \theta_i^2 - \frac{\theta_0^2}{n} &= \sum_{i=1}^{n} \theta_i^2 + \frac{2\theta_0}{n}\sum_{i=1}^{n} \theta_i + \frac{\theta_0^2}{n} \\ &= \sum_{i=1}^{n} \left(\theta_i + \frac{\theta_0}{n}\right)^2 \geq 0,\end{aligned}$$

$$\frac{n+1}{n}\theta_0^2 \leq \sum_{i=0}^{n} \theta_i^2 \leq 1,$$

and so

$$\beta_0 - \alpha_0 = \theta_0 \leq \left(\frac{n}{n+1}\right)^{\frac{1}{2}}.$$

Therefore, we get

$$\begin{aligned}d' + d^* = \sqrt{2}\langle b - a, e\rangle &= \sqrt{2}\sum_{i=0}^{n}(\beta_i - \alpha_i) \\ &\leq (2n(n+1))^{\frac{1}{2}},\end{aligned}$$

and so

$$\min\{d', d^*\} \leq \left(\frac{n(n+1)}{2}\right)^{\frac{1}{2}}.$$

With this, the proof of the lemma is complete. □

Lemma 1.2. *Every 3-dimensional set of diameter 1 can be embedded in a regular octahedron whose opposite facets are a distance 1 apart.*

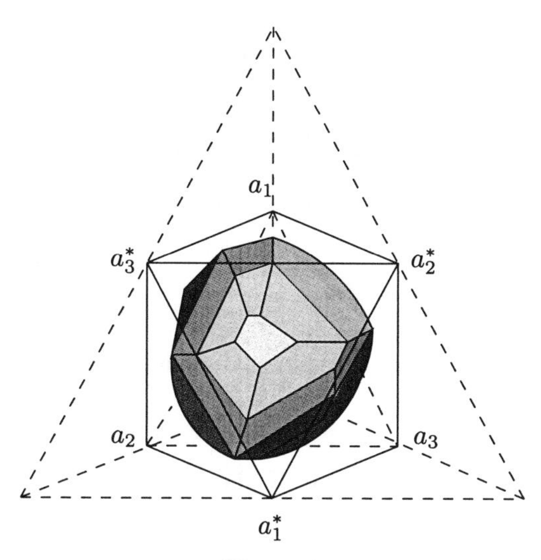

Figure 1

Upon inspection of Figure 1, the interested reader can see that this lemma is an immediate consequence of the previous one.

Proof of Theorem 1.1 (B. Grünbaum [1]): Without loss of generality, assume that $d(X) = 1$. It follows from Lemma 1.2 that the set X can be embedded in a regular octahedron O whose opposite facets are a distance 1 apart. Starting from this octahedron, with vertices a_1, a_2, a_3, a_1^*, a_2^*, and a_3^* such that $a_1a_2a_3$ is a facet of O and a_i^* is the vertex opposite to a_i, we first obtain a better *polyhedron* P such that

$$X \subseteq P \subseteq O. \tag{1.2}$$

Let H_k be the plane containing a_i, a_j, a_i^*, and a_j^*, where $\{i, j, k\} = \{1, 2, 3\}$. The two planes that are parallel to and at a distance of $\frac{1}{2}$ from H_k intersect and cut off from the octahedron two square *pyramids* with a_k and a_k^* as vertices. Clearly, at least one of these two pyramids has no interior point in common with X, and without loss of generality, we assume this pyramid to be the one with a_k^* as a vertex. Since k runs through the values 1, 2, and 3, we are able to cut off all the vertices a_1^*, a_2^*, and a_3^* and replace them by squares $b_ib_i^*c_i^*c_i$ $(i = 1, 2, 3)$, where b_i and b_i^* are on the original facet $a_1^*a_2^*a_3^*$. Thus, we get a polyhedron P which satisfies (1.2). The facet of P contained in $a_1^*a_2^*a_3^*$ is a hexagon with vertices b_1, b_1^*, b_2, b_2^*, b_3, and b_3^*, angles $\frac{2}{3}\pi$, and side lengths

$$\|b_i - b_i^*\| = \frac{\sqrt{3}-1}{\sqrt{2}} \quad \text{and} \quad \|b_i^* - b_{i+1}\| = \frac{2-\sqrt{3}}{\sqrt{2}}$$

(taking $b_4 = b_1$). Its other facets, besides the base $a_1a_2a_3$ and the squares $b_ib_i^*c_i^*c_i$, are three *pentagons* congruent to $b_1^*b_2c_2a_3c_1^*$ (the angle at a_3 is $\frac{1}{3}\pi$, the others are $\frac{2}{3}\pi$; $\|a_3 - c_1^*\| = \|a_3 - c_2\| = \frac{1}{\sqrt{2}}$) and three *trapezoids* congrent to $c_1c_1^*a_3a_2$ (the angles at a_2 and a_3 are $\frac{1}{3}\pi$, the others are $\frac{2}{3}\pi$). As for the diameter, we have

$$d(P) = \|a_i - b_i\| = \|a_i - b_i^*\| = \|a_i - c_i\| = \|a_i - c_i^*\| = \sqrt{2}.$$

To finish, it suffices to show that P can be divided into four parts, P_1, P_2, P_3, and P_4, such that

$$d(P_i) < 1, \quad i = 1, 2, 3, 4. \tag{1.3}$$

For this, it in turn suffices to divide $\partial(P)$ into four parts, Q_1, Q_2, Q_3, and Q_4, such that

$$d(Q_i) < 1, \quad i = 1, 2, 3, 4. \tag{1.4}$$

To see this, note that by Lemma 1.2 there exists a positive number $r \leq \frac{\sqrt{3}}{2} < 1$ and a point z such that $P \subseteq z + rB$. Then (1.4) implies that

$$d(\text{conv}\{z, Q_i\}) < 1, \quad i = 1, 2, 3, 4.$$

Thus, (1.3) will hold if we set $P_i = \text{conv}\{z, Q_i\}$.

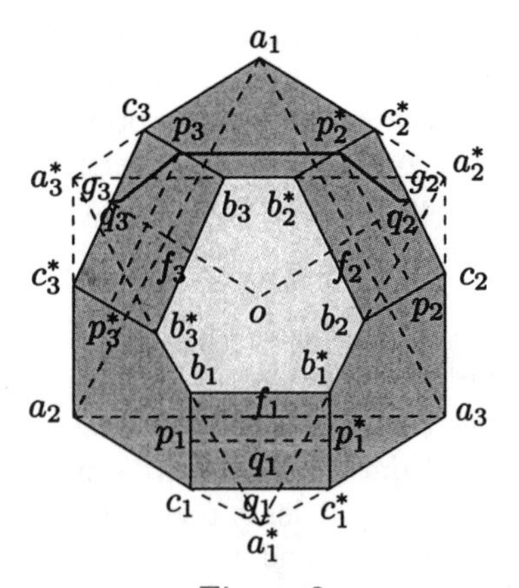

Figure 2

Let o be the center of $a_1a_2a_3$, f_k the midpoint of a_ia_j ($\{i,j,k\} = \{1,2,3\}$), g_i the midpoint of $c_ic_i^*$, p_i and p_i^* the points on the segments b_ic_i and $b_i^*c_i^*$, respectively, such that

$$\|c_i - p_i\| = \|c_i^* - p_i^*\| = \frac{15\sqrt{3} - 10}{46\sqrt{2}}, \quad i = 1, 2, 3,$$

and q_i the point on the segment joining g_i with the midpoint of $b_i b_i^*$ such that

$$\|q_i - g_i\| = \frac{1231\sqrt{3} - 1986}{1518\sqrt{2}}, \quad i = 1, 2, 3.$$

Let Q_1 be the part of $\partial(P)$ which has a_1 as a relative interior point and the closed polygon $of_2g_2q_2p_2^*p_3q_3g_3f_3o$ as its boundary, Q_2 the part of $\partial(P)$ which has a_2 as a relative interior point and the closed polygon $of_3g_3q_3p_3^*p_1q_1g_1f_1o$ as its boundary, Q_3 the part of $\partial(P)$ which has a_3 as a relative interior point and the closed polygon $of_1g_1q_1p_1^*p_2q_2g_2f_2o$ as its boundary, and Q_4 the part of $\partial(P)$ which contains $b_1b_1^*b_2b_2^*b_3b_3^*$ and has $p_1q_1p_1^*p_2q_2p_2^*p_3q_3p_3^*p_1$ as its boundary (see Figure 2). Then, by computing the distances between all pairs of vertices of each Q_i, it can be shown that

$$d(Q_i) < 0.99, \quad i = 1,\ 2,\ 3,\ 4.$$

Thus, Theorem 1.1 is proven. □

§3. Some Remarks

For convenience, we denote by $b(X)$ the smallest number m such that X can be partitioned into parts $X_1, X_2, \ldots, X_m$ such that

$$d(X_i) < d(X), \quad i = 1,\ 2, \ldots, m.$$

In 1932, K. Borsuk [1] proved that

$$b(B) = n + 1$$

by applying the following result:

"*If $\partial(B)$ is covered by the union of n closed sets, then at least one of these sets contains a pair of antipodal points.*"

The 2-dimensional case of Borsuk's problem was solved by K. Borsuk [2] himself. The key to Borsuk's original proof is the following result of J. Pál [1]:

"*Every 2-dimensional set X with $d(X) = 1$ can be inscribed in a regular hexagon whose opposite sides are a distance 1 apart.*"

So far, for dimension greater than 2, an affirmative solution to Borsuk's problem involving no restrictions on the set X is known only for dimension 3. This result was first proved by J. Perkal [1] and H.G. Eggleston [1]. Later, simpler proofs were achieved by B. Grünbaum [1], A. Heppes and P. Révész [1], and A. Heppes [1].

In attacking Borsuk's problem in higher dimensions, many partial results have been achieved. In 1945, H. Hadwiger [1] found that $b(K) \leq n + 1$ for every n-dimensional *smooth convex body* K. In 1955, H. Lenz [1] showed that, on the one hand, $b(K) \leq n$ if K has a smooth boundary but non-constant width, whereas on the other hand, $b(K) \geq n+1$ for all *sets of constant width.* In 1971, A.S. Riesling [1] showed that $b(K) \leq n+1$ for every n-dimensional centrally symmetric convex body K. In 1971, C.A. Rogers [2] proved that $b(K) \leq n+1$ when K is invariant under the *symmetry group* of the regular n-dimensional simplex. Nevertheless, as will be shown in Section 5, the answer to Borsuk's problem is "no" for sufficiently large n. This surprising discovery was made by J. Kahn and G. Kalai [1].

Several upper estimates of $b(X)$ depending only on the dimension n of the set X are known. In 1961, L. Danzer [1] showed that

$$b(X) < \sqrt{\frac{(n+2)^3(2+\sqrt{2})^{n-1}}{3}}.$$

In 1982, M. Lassak [1] proved that

$$b(X) \leq 2^{n-1} + 1.$$

More recently, O. Schramm [1] was able to improve these upper bounds to

$$b(X) \leq 5n^{\frac{3}{2}}(4+\log n)\left(\frac{3}{2}\right)^{\frac{n}{2}}.$$

§4. Larman's Problem

In 1981, D. Larman [1] raised the following combinatorial problem:

Larman's Problem: *Let $\mathcal{A}$ be a family of subsets of $\{1,2,\ldots,n\}$ such that every two members of $\mathcal{A}$ overlap in at least k elements. Can $\mathcal{A}$ be divided into n subfamilies, $\mathcal{A}_1$, $\mathcal{A}_2,\ldots,\mathcal{A}_n$, such that every two members of $\mathcal{A}_i$ overlap in at least $k+1$ elements?*

Given $\mathcal{A}$ as in the statement of Larman's problem, let $l(\mathcal{A},n,k)$ denote the smallest number m for which there exist subfamilies $\mathcal{A}_1$, $\mathcal{A}_2,\ldots,\mathcal{A}_m$ such that

$$\mathcal{A} = \bigcup_{i=1}^{m} \mathcal{A}_i$$

and every two members of $\mathcal{A}_i$ overlap in at least $k+1$ elements. Then an affirmative answer to Larman's problem for the integer n implies that $l(\mathcal{A},n,k) \leq n$.

At first glance, it is not easy to notice a connection between Borsuk's problem and Larman's problem. However, they are closely related. Assume in what follows that every member of the family $\mathcal{A}$ has *cardinality* h.

Denote by T_n the *mapping* from $\mathcal{A}$ to R^n defined by

$$T_n(A) = (x^1, x^2, \ldots, x^n),$$

where

$$x^i = \begin{cases} 0, & i \notin A \\ 1, & i \in A. \end{cases}$$

Note that two members A and A' of $\mathcal{A}$ overlap in exactly j elements if and only if

$$\|T_n(A) - T_n(A')\| = \sqrt{2(h-j)}.$$

Consequently, letting $T_n(\mathcal{A}') = \{T_n(A) : A \in \mathcal{A}'\}$ for any subfamily $\mathcal{A}'$ of $\mathcal{A}$, we see that

$$d(T_n(\mathcal{A}')) \leq \sqrt{2(h-k)}$$

with equality occurring if and only if some two members of $\mathcal{A}'$ overlap in exactly k elements. But then

$$b(T_n(\mathcal{A})) = l(\mathcal{A}, n, k),$$

and so an affirmative answer to Borsuk's problem in dimension n implies that $l(\mathcal{A}, n, k) \leq n+1$.

Clearly now, if for a given n we can find $\mathcal{A}$ and k as above such that

$$l(\mathcal{A}, n, k) > n+1,$$

then the answer to both Larman's problem for the integer n and Borsuk's problem for dimension n will be "no." This is just the starting point of Kahn and Kalai's work.

§5. The Kahn-Kalai Phenomenon

J. Kahn and G. Kalai [1] recently obtained the following unexpected and counterintuitive result.

The Kahn-Kalai Phenomenon: *For every integer $n \geq 1$, there exists a subset X_n of n-dimensional Euclidean space such that*

$$b(X_n) > 1.07^{\sqrt{n}}.$$

This result yields a negative answer to Borsuk's problem for all n large enough so that $1.07^{\sqrt{n}} > n+1$, i.e., for all $n > 21,800$.

The Kahn-Kalai phenomenon will be proven with the help of the following.

Assertion: *For each prime number p, there exists a family $\mathcal{A}_p$ of $4p^2$-element subsets of $\{1, 2, \ldots, 2p(4p-1)\}$ such that every two members of $\mathcal{A}_p$ overlap in at least $2p^2$ elements and*

$$l(\mathcal{A}_p, 2p(4p-1), 2p^2) > 1.5^p. \tag{1.5}$$

Verification of Kahn-Kalai from the Assertion: For $1 \leq n < 2 \cdot 2 \cdot (4 \cdot 2 - 1) = 28$, taking X_n to be a regular n-dimensional simplex, we trivially have that $b(X_n) = n + 1 > 1.07^{\sqrt{n}}$.

For $n \geq 2 \cdot 2 \cdot (4 \cdot 2 - 1) = 28$, let p be the largest prime number such that $2p(4p-1) \leq n$. Letting $\mathcal{A}_p$ be as in the Assertion, we may clearly view $\mathcal{A}_p$ as a family of subsets of $\{1, 2, \ldots, n\}$ and so take $X_n = T_n(\mathcal{A}_p)$. Then by the discussion of the last section,

$$b(X_n) = l(\mathcal{A}_p, n, 2p^2) = l(\mathcal{A}_p, 2p(4p-1), 2p^2) > 1.5^p.$$

A result proved by P. Tchebysheff in 1850 states that for every natural number $m > 1$, there exists a prime number greater than m and less than $2m$ (see page 137 of W. Sierpiński [1]). Thus, letting p' denote the next prime after p, we have that $p' < 2p$. Clearly, then $n < 2p'(4p'-1) < 8p'^2 < 32p^2$. It follows that

$$b(X_n) > 1.5^{\sqrt{n/32}} > 1.07^{\sqrt{n}}$$

and so we are done. □

Note that when $n = 2p(4p-1)$, p a prime, the above shows us that there exists a subset X_n of n-dimensional Euclidean space such that $b(X_n) > 1.5^p$, which is a stronger lower estimate on $b(X_n)$ than that in Kahn-Kalai, which applies to all n. In consequence, we obtain a smaller n for which we are sure that Borsuk's problem has a negative answer, namely $n = 4,186$ corresponding to $p = 23$.

Three lemmas are needed to be able to prove the Assertion.

Lemma 1.3. *For p a prime and l an integer such that $0 \leq l \leq 2p-1$, $\binom{l}{p-1} \equiv 0 \pmod{p}$ if and only if $l \neq p-1$ and $2p-1$.*

Since the proof of this lemma is routine, we omit it.

The following two lemmas and their proofs are both special simple cases of more general results from P. Frankl and R.M. Wilson [1].

Lemma 1.4. *Let p be a prime and $\mathcal{F}$ be a family of $(2p-1)$-element subsets of $\{1, 2, \ldots, n\}$ such that*

$$\operatorname{card}\{F \cap F'\} \neq p - 1$$

for every two distinct members $F, F' \in \mathcal{F}$. Then

$$\operatorname{card}\{\mathcal{F}\} \leq \binom{n}{p-1}.$$

Proof: Let F_1', $F_2', \ldots, F'_{\binom{n}{p-1}}$ be all the $(p-1)$-element subsets and F_1, F_2, $\ldots, F_{\binom{n}{2p-1}}$ be all the $(2p-1)$-element subsets of $\{1, 2, \ldots, n\}$. Let M be the $\binom{n}{p-1} \times \binom{n}{2p-1}$ matrix whose (u, v)-entry is

$$m_{u,v} = \begin{cases} 1, & F_u' \subset F_v \\ 0, & F_u' \not\subset F_v, \end{cases} \tag{1.6}$$

where $1 \leq u \leq \binom{n}{p-1}$ and $1 \leq v \leq \binom{n}{2p-1}$.

Let V denote the vector space generated by the row vectors of M. Clearly,

$$\dim\{V\} \leq \binom{n}{p-1}. \tag{1.7}$$

Setting $M^* = M^T M$ where M^T is the transpose of M, it is easy to see that M^* is an $\binom{n}{2p-1} \times \binom{n}{2p-1}$ matrix whose row vectors are contained in V. Obviously, then

$$\text{rank}\{M^*\} \leq \dim\{V\}. \tag{1.8}$$

If $M^*_{\mathcal{F}}$ is the $\text{card}\{\mathcal{F}\} \times \text{card}\{\mathcal{F}\}$ submatrix of M^* corresponding to $\mathcal{F}$, then trivially

$$\text{rank}\{M^*_{\mathcal{F}}\} \leq \text{rank}\{M^*\}. \tag{1.9}$$

Using (1.6), it can be shown that the (u, v)-entry of M^* is

$$m^*_{u,v} = \binom{\text{card}\{F_u \cap F_v\}}{p-1}, \quad 1 \leq u, v \leq \binom{n}{2p-1}. \tag{1.10}$$

It follows from (1.10), the lemma's hypothesis, and Lemma 1.3, that

$$m^*_{u,v} \equiv 0 \pmod{p}, \quad F_u, F_v \in \mathcal{F}, \ u \neq v \tag{1.11}$$

and

$$m^*_{u,u} \not\equiv 0 \pmod{p}, \quad F_u \in \mathcal{F}. \tag{1.12}$$

Because of (1.11) and (1.12), $\det\{M^*_{\mathcal{F}}\} \not\equiv 0 \pmod{p}$. But then $M^*_{\mathcal{F}}$ is invertible, and so

$$\text{card}\{\mathcal{F}\} = \text{rank}\{M^*_{\mathcal{F}}\}. \tag{1.13}$$

Stringing together (1.7), (1.8), (1.9), and (1.13), we get that

$$\text{card}\{\mathcal{F}\} \leq \binom{n}{p-1},$$

which proves our lemma. □

Lemma 1.5. *For p a prime, let $m(p)$ be the maximum number of $2p$-element subsets of $\{1, 2, \ldots, 4p\}$ such that no two of them overlap in p elements. Then*

$$m(p) \leq \frac{1}{2}\binom{4p}{p}. \tag{1.14}$$

Proof: Assume that $\mathcal{F}$ is a family of $2p$-element subsets of $\{1,2,\ldots,4p\}$, no two of which overlap in p elements, and that $\text{card}\{\mathcal{F}\} = m(p)$. For $1 \le i \le 4p$, let $\mathcal{F}_i$ be the collection of those members of $\mathcal{F}$ which contain i. Of course,

$$\sum_{i=1}^{4p} \text{card}\{\mathcal{F}_i\} = 2p \cdot m(p).$$

Assuming, without loss of generality, that $\text{card}\{\mathcal{F}_{4p}\}$ is maximal among all $\text{card}\{\mathcal{F}_i\}$, we see that

$$\sum_{i=1}^{4p} \text{card}\{\mathcal{F}_i\} \le 4p \cdot \text{card}\{\mathcal{F}_{4p}\},$$

which in conjunction with our last displayed formula immediately yields

$$\text{card}\{\mathcal{F}_{4p}\} \ge \tfrac{1}{2} m(p). \tag{1.15}$$

Setting

$$\mathcal{G} = \{F \setminus \{4p\} : \; F \in \mathcal{F}_{4p}\},$$

we see that $\mathcal{G}$ is a family of $(2p-1)$-element subsets of $\{1,2,\ldots,4p-1\}$, no two of which overlap in $p-1$ elements. Applying Lemma 1.4, we have that

$$\text{card}\{\mathcal{F}_{4p}\} = \text{card}\{\mathcal{G}\} \le \binom{4p-1}{p-1}. \tag{1.16}$$

Together (1.15) and (1.16) show that

$$m(p) \le 2\binom{4p-1}{p-1} = \frac{1}{2}\binom{4p}{p}.$$

Thus Lemma 1.5 is proven. □

Verification of the Assertion: Recall that, given a prime p, it suffices to produce a family $\mathcal{A}_p$ of $4p^2$-element subsets of $\{1,2,\ldots,2p(4p-1)\}$ such that every two members of $\mathcal{A}_p$ overlap in at least $2p^2$ elements, and then to show that (1.5) holds for this $\mathcal{A}_p$. Clearly, the base set $\{1,2,\ldots,2p(4p-1)\}$ can be replaced by any set M of cardinality $2p(4p-1)$. We take M to be the set of all pairs of elements from $S = \{1,2,\ldots,4p\}$ since our family $\mathcal{A}_p$ is most easily defined as a collection of subsets of this M.

For $A \subseteq S$, let $P(A)$ be the set of all pairs which contain one element of A and one element of $S \setminus A$. Define $\mathcal{A}_p$ to be the family of all such sets of pairs determined by those subsets of S which split S up into two equal parts, i.e.,

$$\mathcal{A}_p = \{P(A) : A \subseteq S \text{ and } \text{card}\{A\} = 2p\}.$$

Given A and $A' \subseteq S$, if $\operatorname{card}\{A \cap A'\} = r$, then $\operatorname{card}\{A \cap (S \setminus A')\} = \operatorname{card}\{(S \setminus A) \cap A'\} = 2p - r$ and $\operatorname{card}\{(S \setminus A) \cap (S \setminus A')\} = r$. Consequently,

$$\operatorname{card}\{P(A) \cap P(A')\} = r^2 + (2p - r)^2 = 2(r - p)^2 + 2p^2. \tag{1.17}$$

Note that every member of $\mathcal{A}_p$ obviously has $4p^2$ elements, and now every two members of $\mathcal{A}_p$ clearly overlap, by (1.17), in at least $2p^2$ elements.

All that remains is to verify (1.5) for $\mathcal{A}_p$. Denote $l(\mathcal{A}_p, 2p(4p-1), 2p^2)$ more briefly by m. Then, by definition there exist subfamilies $\mathcal{A}_1, \mathcal{A}_2, \ldots, \mathcal{A}_m$ whose union is all of $\mathcal{A}_p$ such that every two members of $\mathcal{A}_i$ overlap in at least $2p^2 + 1$ elements. We may write each $\mathcal{A}_i$ as $\{P(A) : A \in \mathcal{F}_i\}$, where each $\mathcal{F}_i$ is a family of $2p$-element subsets of the $4p$-element set S.

From (1.17) it is easy to deduce that $P(A) = P(A')$ if and only if $A' = A$ or $A' = S \setminus A$, so the correspondence $A \to P(A)$ is exactly two-to-one. In consequence, the cardinality of $\mathcal{A}_p$ is not $\binom{4p}{2p}$, but rather $\binom{4p}{2p}/2$. Hence,

$$\frac{1}{2}\binom{4p}{2p} \leq \sum_{i=1}^{m} \operatorname{card}\{\mathcal{A}_i\} \leq \sum_{i=1}^{m} \operatorname{card}\{\mathcal{F}_i\}. \tag{1.18}$$

From (1.17) it is also easy to deduce that $\operatorname{card}\{P(A) \cap P(A')\} \geq 2p^2 + 1$ if and only if $\operatorname{card}\{A \cap A'\} \neq p$. This implies that no two members of $\mathcal{F}_i$ overlap in p elements; thus by Lemma 1.5, $\operatorname{card}\{\mathcal{F}_i\} \leq \frac{1}{2}\binom{4p}{p}$. Hence,

$$\sum_{i=1}^{m} \operatorname{card}\{\mathcal{F}_i\} \leq \frac{m}{2}\binom{4p}{p}. \tag{1.19}$$

From (1.18) and (1.19) one immediately sees that

$$\begin{aligned} m &\geq \frac{\binom{4p}{2p}}{\binom{4p}{p}} = \frac{3p(3p-1)\cdots(3p-(p-1))}{2p(2p-1)\cdots(2p-(p-1))} \\ &= \frac{3}{2} \cdot \frac{3 - 1/p}{2 - 1/p} \cdot \cdots \cdot \frac{3 - (p-1)/p}{2 - (p-1)/p} > \left(\frac{3}{2}\right)^p. \end{aligned}$$

Since this is (1.5), we are done. □

2

Finite Packing Problems

§1. Introduction

In n-dimensional Euclidean space, how should one arrange m nonoverlapping translates of a given convex body K in order to minimize the diameter, the surface area, or the volume of their convex hull?

In 1975, L. Fejes Tóth [4] studied the case of this problem involving volume. He noticed that when $n \geq 7$, a linear arrangement of m unit balls yields a greater *local density* than the *densest lattice packing.* This led him to raise the following well-known conjecture.

The Sausage Conjecture: *In n-dimensional Euclidean space with $n \geq 5$, the volume of the convex hull of m nonoverlapping unit balls is at least $2(m-1)\omega_{n-1} + \omega_n$, with equality being attained only when the centers of these balls are equally spaced a distance 2 apart on a line.*

In contrast, H.T. Croft, K.J. Falconer, and R.K. Guy [1, pp. 117–118] have formulated the following counterpart concerning surface area and diameter.

The Spherical Conjecture: *A convex body of minimal surface area, or alternately of minimal diameter, into which m unit balls can be packed is roughly spherical when m is large.*

At first glance, it is really hard to imagine that both of these conjectures can hold simultaneously. However, recently, the sausage conjecture has been proven by U. Betke, M. Henk, and J.M. Wills [1] for n sufficiently large, whereas the spherical conjecture has been confirmed by K. Böröczky Jr. [2] and C. Zong [3]. It is with these results that this chapter is concerned.

§2. Supporting Functions, Area Functions, Minkowski Sums, Mixed Volumes, and Quermassintegrals

The notions of *supporting function, area function, Minkowski sum, mixed volume*, and *quermassintegral* are five fundamental concepts in Convex and Discrete Geometry. This section is devoted to recalling their definitions and stating those of their basic properties which will be needed in this book.

Supporting function: *The supporting function h_K of a given convex body K is defined by*

$$h_K(x) = \max\{\langle x, y\rangle : \; y \in K\}.$$

Area function: *Given U, a Borel subset of $\partial(B)$, let $K(U)$ denote the set of those points of $\partial(K)$ at which there exists a supporting hyperplane of K with an external unit normal $u \in U$. The area function of K is the measure $U \mapsto G(K,U)$ on $\partial(B)$ defined by*

$$G(K,U) = s(K(U)).$$

Here, $s(*)$ denotes the surface area measure on $\partial(K)$.

Minkowski sum: *If K and L are convex bodies, then*

$$K + L = \{x + y : \; x \in K, \; y \in L\}$$

is called the Minkowski sum of K and L.

Mixed volume and quermassintegral: *For K and L two given convex bodies in R^n, it can be shown that the function $\rho \mapsto v(K + \rho L)$ is a polynomial of degree n. Consequently, there are uniquely determined numbers $V_i(K, L)$ for $0 \le i \le n$ such that*

$$v(K + \rho L) = \sum_{i=0}^{n} V_i(K,L)\binom{n}{i}\rho^i. \tag{2.1}$$

Clearly, $V_0(K,L) = v(K)$ and $V_n(K,L) = v(L)$. The coefficient $V_i(K,L)$ is called the i-th mixed volume of K and L. We call $W_i(K) = V_i(K,B)$ the i-th quermassintegral of K.

The following remarks contain the basic results about these concepts that are necessary to what follows.

Remark 2.1. *The volume of K can be expressed as*

$$v(K) = \frac{1}{n} \int_{\partial(B)} h_K(u) G(K, du).$$

Remark 2.2. *$W_i(K)$ is invariant under rigid transformations of K.*

Remark 2.3. *For any non-negative number λ, we have*

$$W_i(\lambda K) = \lambda^{n-i}\, W_i(K).$$

Remark 2.4. *$W_i(K)$ depends continuously on K under the Hausdorff metric.*

Remark 2.5. *If $K \subseteq L$, then $W_i(K) \leq W_i(L)$.*

Remark 2.6. *For any convex body K, $W_i(K) \geq 0$, with equality being attained only when* $\dim\{K\} < n - i$.

Remark 2.7. *The volume, surface area, and diameter of a convex body K are related to its quermassintegrals as follows:*

$$v(K) = W_0(K), \quad s(K) = nW_1(K)$$

and

$$d(K) \geq \frac{2}{\omega_n} W_{n-1}(K).$$

Besides these simple properties, the following lemmas will play very important roles in the proofs of Theorem 2.1, the *Venkov-McMullen Theorem*, and *Schneider's phenomenon* to follow. Since these lemmas are well known and their proofs are complicated, we state the lemmas only. The omitted proofs can be found in K. Leichtweiß [2], or, for an English language source, H.G. Eggleston [2].

Lemma 2.1 (The Brunn-Minkowski inequality). *Given K and L arbitrary convex bodies and θ any real number strictly between* 0 *and* 1, *we have*

$$\sqrt[n]{v\left((1-\theta)K + \theta L\right)} \geq (1-\theta)\sqrt[n]{v(K)} + \theta \sqrt[n]{v(L)},$$

with equality being attained only when K and L are homothetic, i.e., only when $K = z + \lambda L$ for a suitable point z and a suitable number λ.

Lemma 2.2. *The first mixed volume of K and L satisfies both*

$$V_1(K,L) = \frac{1}{n}\int_{\partial(B)} h_L(u)G(K,du) \tag{2.2}$$

and[1]

$$V_1(K,L) \geq v(K)^{\frac{n-1}{n}} v(L)^{\frac{1}{n}}. \tag{2.3}$$

Corollary 2.1. *It follows easily from* (2.2) *that for any unit vector u,*

$$v(P_u(K)) = \frac{1}{2}\int_{\partial(B)} |\langle u,w\rangle| G(K,dw)$$

where P_u denotes the orthogonal projection onto the hyperplane $H_u = \{x \in R^n : \langle x,u\rangle = 0\}$.

Lemma 2.3 (The isoperimetric inequality). *Let i be such that $1 \leq i \leq n-1$. Then as K varies over all n-dimensional convex bodies such that $v(K) = \omega_n$, the minimum value of $W_i(K)$ is attained only when K is the unit ball.*

§3. The Optimal Finite Packings Regarding Quermassintegrals

It follows from Remark 2.7 that the following general result of K. Böröczky Jr. [2] and C. Zong [3] has, as a particular consequence, an affirmative answer to the spherical conjecture of H.T. Croft, K.J. Falconer, and R.K. Guy.

Theorem 2.1 (K. Böröczky Jr. [2] and C. Zong [3]). *For every n-dimensional convex body K and every index i with $1 \leq i \leq n-1$, let $\mathcal{K}_{i,m}$ be the family of all convex bodies $K_{i,m}$ which contain m nonoverlapping translates of K and at which the i-th quermassintegral is minimum. Define*

$$\eta_{i,m} = \max_{K_{i,m}\in\mathcal{K}_{i,m}} \left\{ \min_{\substack{x+rB\subseteq K_{i,m}\subseteq x+r'B \\ x\in R^n}} \frac{r'}{r} \right\}.$$

Then

$$\lim_{m\to+\infty} \eta_{i,m} = 1.$$

To prove this theorem, we need, in addition to Lemma 2.3, the following results:

[1] Inequality (2.3) is a special case of the famous *Aleksandrov-Fenchel inequality.*

Lemma 2.4. *For every n-dimensional convex body K, there exists a rectangular parallelepiped P centered at the origin and a suitable point z such that*

$$z + P \subseteq K \subseteq z + n^{\frac{3}{2}} P.$$

Proof: The proof is divided into three parts:

Assertion 2.1 (K. Leichtweiß [1]). *Let α and β be two fixed numbers with $|\alpha| \leq 1$ and $|\beta| \leq 1$. If for the ellipsoids*

$$E_\lambda : \quad \lambda(x^1 - \alpha)(x^1 + \beta) + \sum_{i=1}^{n} (x^i)^2 \leq 1$$

we have that $v(E_\lambda) \geq \omega_n$ whenever $0 \leq \lambda < +\infty$, then $\alpha\beta \geq \frac{1}{n}$.

Recalling that the volume of an n-dimensional ellipsoid is simply the product of the lengths of its semiaxes and ω_n, a little algebra easily shows that

$$v(E_\lambda) = \left(1 + (1 + \alpha\beta)\lambda + \frac{(\alpha+\beta)^2}{4}\lambda^2\right)^{\frac{n}{2}} (1+\lambda)^{-\frac{n+1}{2}} \omega_n.$$

Hence,

$$\left.\frac{d(v(E_\lambda))}{d\lambda}\right|_{\lambda=0} = \frac{1}{2}(n\alpha\beta - 1)\omega_n.$$

But then our assumption that $v(E_\lambda) \geq \omega_n = v(E_0)$ for all $0 \leq \lambda < +\infty$ implies that

$$\left.\frac{d(v(E_\lambda))}{d\lambda}\right|_{\lambda=0} \geq 0,$$

and therefore

$$\alpha\beta \geq \frac{1}{n}.$$

Assertion 2.2 (K. Leichtweiß [1]). *For every n-dimensional convex body K, there is an ellipsoid E centered at the origin and a suitable point z such that*

$$z + \frac{1}{n} E \subseteq K \subseteq z + E.$$

Among all ellipsoids E centered at the origin and all points z such that $K \subseteq z + E$, let $z + E$ be chosen with minimal volume. Without loss of generality, by translating we may assume that z is the origin. Let T be a linear transformation of R^n such that $T(E) = B$. Clearly, B is an ellipsoid containing $T(K)$ of minimal volume among all translated ellipsoids containing $T(K)$. Let H_1 and H_2 be two parallel supporting hyperplanes of $T(K)$. Without loss of generality, by rotating we may assume that

$$H_1 = \{(x^1, x^2, \ldots, x^n) : x^1 = \alpha\}$$

and

$$H_2 = \{(x^1, x^2, \dots, x^n) : x^2 = -\beta\}.$$

We may also assume, since $T(K) \subseteq B$, that $-1 \leq -\beta \leq \alpha \leq 1$. It is easy to see that every point $(x^1, x^2, \dots, x^n) \in T(K)$ satisfies both

$$\sum_{i=1}^{n} (x^i)^2 \leq 1 \quad \text{and} \quad (x^1 - \alpha)(x^1 + \beta) \leq 0.$$

Thus, we get that

$$T(K) \subseteq E_\lambda$$

and hence, by the minimality of the volume of B, that $v(E_\lambda) \geq \omega_n$ for all $0 \leq \lambda < +\infty$. By Assertion 2.1, $\alpha\beta \geq n^{-1}$. This forces both α and β to be > 0 and hence $\geq n^{-1}$. Consequently, $n^{-1}B$ lies between H_1 and H_2. Since any convex set is the intersection of all the regions between all its pairs of parallel supporting hyperplanes, we conclude that

$$\frac{1}{n} B \subseteq T(K) \subseteq B$$

and therefore that

$$z + \frac{1}{n} E \subseteq K \subseteq z + E.$$

Now we turn to the last step. Denote the semiaxes of the ellipsoid E in Assertion 2.2 by $a_1 u_1$, $a_2 u_2, \dots$, and $a_n u_n$, where the a_i's are positive scalars and the u_i's form an orthonormal set of vectors. Setting

$$P = \{x \in R^n : |\langle x, u_i \rangle| \leq n^{-\frac{3}{2}} a_i,\ i = 1, 2, \dots, n\}$$

and letting z be as in Assertion 2.2, we get

$$z + P \subseteq z + \frac{1}{n} E \subseteq K \subseteq z + E \subseteq z + n^{\frac{3}{2}} P.$$

Lemma 2.4 is proven. □

Lemma 2.5 (E. Hlawka [1]). *For a given convex body K, let $m(M, K)$ be the maximum number of translates $x + K$ which can be packed into a convex body M. Then*

$$\lim_{\rho \to +\infty} \frac{m(\rho M, K) v(K)}{v(\rho M)} = \delta(K)$$

exists and does not depend on M. Usually, $\delta(K)$ is called the density of the densest translative packings of K.

Proof: Let W be the unit cube $\{(x^1, x^2, \dots, x^n) : |x^i| \leq \frac{1}{2},\ i = 1, 2, \dots, n\}$ and let $m^*(M, K)$ be the maximum number of the nonoverlapping translates $K + x$ which intersect M. Define

$$\delta(K) = \limsup_{l \to +\infty} \frac{m(lW, K) v(K)}{v(lW)}.$$

Then, on the one hand, since

$$\lim_{l\to+\infty}\frac{v((l+2d(K))W)-v((l-2d(K))W)}{v(lW)}=0,$$

it follows that for each $\epsilon>0$, there is an l_ϵ such that

$$\frac{m(l_\epsilon W,K)v(K)}{v(l_\epsilon W)}>\delta(K)-\epsilon$$

and

$$\frac{m^*(l_\epsilon W,K)v(K)}{v(l_\epsilon W)}<\delta(K)+\epsilon.$$

On the other hand, for each convex body M, there exists a $\rho_\epsilon>0$ such that for each $\rho>\rho_\epsilon$, there are two families of nonoverlapping cubes $\{\frac{l_\epsilon}{\rho}W+x_i:\ i=1,2,\ldots,g\}$ and $\{\frac{l_\epsilon}{\rho}W+y_i:\ i=1,2,\ldots,h\}$ such that

$$\bigcup_{i=1}^{g}\left(\frac{l_\epsilon}{\rho}W+x_i\right)\subseteq M,\qquad v(M)\leq\frac{v\left(\bigcup_{i=1}^{g}\left(\frac{l_\epsilon}{\rho}W+x_i\right)\right)}{1-\epsilon}$$

and

$$M\subseteq\bigcup_{i=1}^{h}\left(\frac{l_\epsilon}{\rho}W+y_i\right),\qquad v(M)\geq\frac{v\left(\bigcup_{i=1}^{h}\left(\frac{l_\epsilon}{\rho}W+y_i\right)\right)}{1+\epsilon}.$$

Hence, when $\rho>\rho_\epsilon$, one has

$$(1-\epsilon)(\delta(K)-\epsilon)<\frac{m(\rho M,K)v(K)}{v(\rho M)}<(1+\epsilon)(\delta(K)+\epsilon),$$

which implies the assertions of Lemma 2.5. □

Proof of Theorem 2.1 (C. Zong [3]): By Lemma 2.4, for every convex body $K_{i,m}\in\mathcal{K}_{i,m}$, there is a rectangular parallelepiped P and a suitable point z such that

$$z+P\subseteq K_{i,m}\subseteq z+n^{\frac{3}{2}}P. \tag{2.4}$$

Let $l_1,\,l_2,\ldots,l_n$ be the lengths of the n edges of P and define

$$\Psi_i=\sum_{j_1,j_2,\ldots,j_i\text{ pairwise distinct}}l_{j_1}l_{j_2}\ldots l_{j_i}.$$

From (2.4) and Remark 2.5 we get

$$mv(K)\leq v(K_{i,m})\leq v(n^{\frac{3}{2}}P)=n^{\frac{3n}{2}}\prod_{j=1}^{n}l_j \tag{2.5}$$

and

$$W_i(K_{i,m}) \geq W_i(P) = \frac{\omega_i}{\binom{n}{i}} \Psi_{n-i}. \tag{2.6}$$

On the other hand, from Lemma 2.5, it is easy to see that m translates of K can be packed into a ball with radius

$$r = 2\left(\frac{mv(K)}{\delta(K)\omega_n}\right)^{\frac{1}{n}}$$

when m is sufficiently large. Then, by the minimality assumption on $W_i(K_{i,m})$, we must have $W_i(K_{i,m}) \leq W_i(rB)$. Hence, keeping (2.5), the fact that $W_i(rB) = \omega_n r^{n-i}$, and (2.6) in mind, we get that

$$\begin{aligned}\frac{\omega_i}{\binom{n}{i}} \Psi_{n-i} &\leq \omega_n 2^{n-i} \left(\frac{mv(K)}{\delta(K)\omega_n}\right)^{\frac{n-i}{n}} \\ &\leq 2^{n-i} n^{\frac{3(n-i)}{2}} (\delta(K))^{\frac{i-n}{n}} (\omega_n)^{\frac{i}{n}} \left(\prod_{i=1}^{n} l_i\right)^{\frac{n-i}{n}}.\end{aligned}$$

This, together with a general lower bound for $\delta(K)$ on page 36 of C.A. Rogers [3], implies that

$$\frac{\Psi_{n-i}}{\left(\prod_{i=1}^n l_i\right)^{\frac{n-i}{n}}} \leq c_1(n,i)$$

and therefore

$$\frac{\max\limits_{1\leq j\leq n}\{l_j\}}{\min\limits_{1\leq j\leq n}\{l_j\}} \leq c_2(n,i), \tag{2.7}$$

where $c_1(n,i)$ and $c_2(n,i)$ are positive numbers depending only on n and i. Then, by (2.4) and (2.7), for every $K_{i,m} \in \mathcal{K}_{i,m}$ there is an $\epsilon(K_{i,m})$ and a suitable point $z(K_{i,m})$ such that

$$z(K_{i,m}) + c_3(n,i)B \subseteq \epsilon(K_{i,m})K_{i,m} \subseteq z(K_{i,m}) + B,$$

where $c_3(n,i) > 0$ depends only on n and i.

If $\lim_{m\to+\infty} \eta_{i,m} \neq 1$, then by *Blaschke's selection theorem*, there is a non-spherical convex body L_i and a series of convex hulls $K_{i,i_1}, K_{i,i_2}, \ldots$ such that, without loss of generality,

$$c_3(n,i)B \subseteq L_i \subseteq B$$

and

$$\lim_{i_m\to+\infty} \epsilon(K_{i,i_m})K_{i,i_m} = L_i.$$

Then, applying Lemma 2.3, we can find three positive numbers r_i, α_i, and β_i such that

$$v(r_iB) \geq v(L_i) + \alpha_i$$

and

$$W_i(r_iB) \leq W_i(L_i) - \beta_i.$$

Applying Lemma 2.5 once more, it can be seen that $\frac{r_i}{\epsilon(K_{i,i_m})}B$ contains i_m nonoverlapping translates of K but has i-th quermassintegral smaller than $W_i(K_{i,i_m})$ when i_m is sufficiently large. This contradiction proves that

$$\lim_{m\to+\infty} \eta_{i,m} = 1$$

and hence completes the proof. □

Remark 2.8. *A similar proof shows that the conclusion of Theorem 2.1 holds even when the m translates are restricted to be taken from an arbitary lattice packing of K.*

§4. The L. Fejes Tóth-Betke-Henk-Wills Phenomenon

Let L_m be a linear segment with length $2(m-1)$ and define

$$L_m^n = L_m + B^n,$$

where the index n indicates the dimension. It is easy to see that L_m^n is the shortest "sausage" which contains m n-dimensional unit balls. By a routine calculation, for m large we get

$$\frac{m\omega_2}{v(L_m^2)} = \frac{m\omega_2}{4(m-1)+\omega_2} < \frac{\pi}{2\sqrt{3}} = \delta(B^2)$$

and

$$\frac{m\omega_3}{v(L_m^3)} = \frac{m\omega_3}{2(m-1)\omega_2+\omega_3} < \frac{\pi}{3\sqrt{2}} \leq \delta(B^3).$$

This means that in dimensions 2 and 3, when m is large, the sausage arrangement does not minimize the volume of a convex hull which contains m unit balls. However U. Betke, M. Henk, and J.M. Wills [1] were recently able confirm the sausage conjecture for very high dimensions. Denote the family of all those convex bodies of minimal volume containing m disjoint unit balls by $\mathcal{B}_{0,m}$.

The L. Fejes Tóth-Betke-Henk-Wills Phenomenon: *Among all the n-dimensional convex bodies B_m which contain m nonoverlapping unit balls,* *the "sausages" L_m^n are the only ones at which the minimum volume is attained when n is large. In other words, the "sausages" L_m^n are the only elements of $\mathcal{B}_{0,m}$ when n is large.*

Assume that $x_i + B$, $i = 1, 2, \ldots, m$, are the m disjoint unit balls in B_m and set $X_m^n = \{x_1, x_2, \ldots, x_m\}$. As usual, we call

$$D_i = \{x \in R^n : \|x - x_i\| \le \|x - x_j\| \text{ for all } x_j \in X_m^n\}$$

the *Dirichlet-Voronoi cell* of X_m^n with respect to x_i. Clearly, we have

$$v(B_m) \ge \sum_{i=1}^{m} v(D_i \cap B_m). \tag{2.8}$$

Thus, to verify the L. Fejes Tóth-Betke-Henk-Wills phenomenon, it is sufficient to show that $v(D_i \cap B_m) \ge 2\omega_{n-1}$ for some $m-2$ Dirichlet-Voronoi cells, that $v(D_i \cap B_m) \ge \omega_{n-1} + \frac{1}{2}\omega_n$ for the two remaining Dirichlet-Voronoi cells, and that equality in these equations cannot hold simultaneously if $B_m \ne L_m^n$. To this end, we consider a fixed Dirichlet-Voronoi cell, say $D = D_m$ with respect to x_m, and set $Q = D \cap B_m$. By translation we may assume that $x_m = o$.

First, we state a basic result concerning this Dirichlet-Voronoi cell.

Lemma 2.6. *The distance from o to the $(n-i)$-dimensional plane determined by any $(n-i)$-dimensional face of D is at least $\sqrt{\frac{2i}{i+1}}$ for $1 \le i \le n$.*

Proof: The $(n-i)$-dimensional plane determined by any $(n-i)$-dimensional face of D is the intersection of at least i $(n-1)$-dimensional planes containing $(n-1)$-dimensional faces of D. Hence, there are i centers, $x_1, x_2, \ldots, x_i$ say, such that the $(n-i)$-dimensional plane is in the intersection of the $(n-1)$-dimensional planes that are the perpendicular bisectors of the segments $ox_1, ox_2, \ldots, ox_i$. Consider any point x on the $(n-i)$-dimensional plane. Since x is equidistant from o and x_1, from o and $x_2, \ldots,$ and from o and x_i, it is clear that x is equidistant from the points $o, x_1, x_2, \ldots, x_i$.

Now write $y_0 = o - x$, $y_1 = x_1 - x$, $y_2 = x_2 - x$, $\ldots$, $y_i = x_i - x$. Since none of the unit balls in the system $X_m^n + B$ overlap, we have

$$\begin{aligned} 2i(i+1) &= \sum_{0 \le k < l \le i} 4 \le \sum_{0 \le k < l \le i} \langle y_k - y_l, y_k - y_l \rangle \\ &= (i+1) \sum_{0 \le k \le i} \langle y_k, y_k \rangle - \left\langle \sum_{0 \le k \le i} y_k, \sum_{0 \le k \le i} y_k \right\rangle \\ &\le (i+1) \sum_{0 \le k \le i} \langle y_k, y_k \rangle = (i+1)^2 \langle y_0, y_0 \rangle. \end{aligned}$$

Hence, we get

$$\|x - o\| \geq \sqrt{\frac{2i}{i+1}},$$

which proves the lemma. □

Definition 2.1. *Let* $y_i = \frac{x_i}{\|x_i - o\|}$ *for* $i = 1, 2, \ldots, m-1$ *and set*

$$\phi = \max_{1 \leq k, l \leq m-1} \{\arccos(|\langle y_k, y_l \rangle|)\},$$

where $\arccos(*)$ *is chosen in* $[0, \frac{\pi}{2}]$. *Further, by relabelling indices if necessary, it may be assumed that* y_1 *and* y_2 *satisfy*

$$\arccos(|\langle y_1, y_2 \rangle|) =$$

$$\begin{cases} \phi, & \text{if } \phi \geq \frac{\pi}{3} \text{ or } \langle y_k, y_l \rangle \geq 0 \text{ for all } \\ & 1 \leq k,\, l \leq m-1; \\ \max\limits_{\langle y_k, y_l \rangle < 0} \{\arccos(|\langle y_k, y_l \rangle|)\}, & \text{in all other cases.} \end{cases} \tag{2.9}$$

Remark 2.9. *In the second case of* (2.9) *we define*

$$V_1 = \left\{ y \in \partial(B) : \ \langle y, y_1 \rangle > \tfrac{1}{2} \right\}$$

and

$$V_2 = \{ y \in \partial(B) : \ \langle y, y_1 \rangle \leq \langle y_2, y_1 \rangle \}.$$

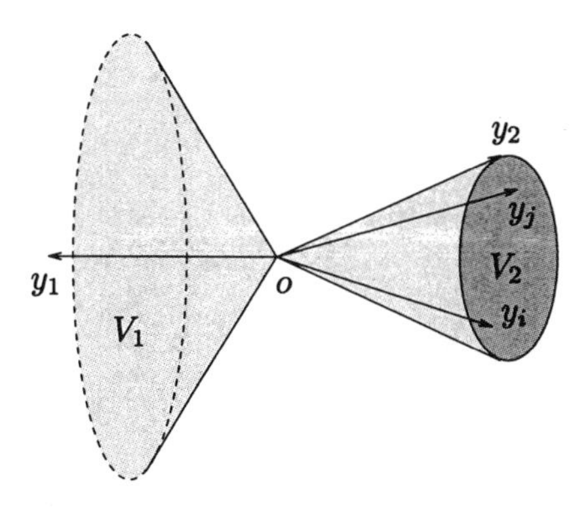

Figure 3

It is easy to see that

$$\{y_1, y_2, \ldots, y_{m-1}\} \subset V_1 \cup V_2$$

and that the angle between any two vectors $u \in V_1$ *and* $v \in V_2$ *is larger than* $\frac{\pi}{3}$. *Hence, a routine argument yields*

$$-\cos\left(\frac{\phi}{2}\right) \le \langle y_1, y_2\rangle \le -\cos(\phi). \tag{2.10}$$

Remark 2.10. *It follows from Definition* 2.1 *that, roughly speaking, when* ϕ *is small, the convex hull of* $X^n_m + B$ *in the vicinity of* o *is "sausage-like" if* $\langle y_1, y_2\rangle < 0$ *and "needle-like" if* $\langle y_1, y_2\rangle > 0$. *Also, since the angle sum of a triangle is* π, *a little thought shows that one cannot have* $\phi < \frac{\pi}{3}$ *and* $\langle y_1, y_2\rangle > 0$ *holding at three points of* X^n_m. *This assertion plays a very important role in the proof of the L. Fejes Tóth-Betke-Henk-Wills phenomenon.*

Definition 2.2. *Set* $G = \mathrm{conv}\{o, 2y_1, 2y_2\} \cap B$ *and let* $p(G,x)$ *denote the nearest point map of* R^n *with respect to* G *given by*

$$p(G,x) = \textit{the unique } y \in G \textit{ such that } \|x-y\| = \min_{z\in G}\{\|x-z\|\}.$$

Then define

$$\begin{aligned}
Q_1 &= \{x \in Q:\ p(G,x) \in \mathrm{int}(G)\},\\
Q_2 &= \{x \in Q:\ p(G,x) \in \mathrm{conv}\{o,y_1\} \cup \mathrm{conv}\{o,y_2\} \setminus \{o,y_1,y_2\}\},\\
Q_3 &= \{x \in Q:\ p(G,x) = o\},\\
Q_4 &= \{x \in Q:\ p(G,x) \in \mathrm{conv}\{2y_1, 2y_2\}\}.
\end{aligned}$$

Clearly, we have

$$v(Q) \ge \sum_{i=1}^{4} v(Q_i). \tag{2.11}$$

The L. Fejes Tóth-Betke-Henk-Wills phenomenon will be a consequence of (2.8), (2.11), and various estimates on the sets $v(Q_i)$ which follow.

Lemma 2.7. *Suppose* $u \in \partial(B)$ *and* $v \in \{x \in \partial(B):\ \langle x,u\rangle = 0\}$. *Let* $\alpha > 0$ *and* $\epsilon > 0$ *be such that* $\alpha(\alpha+\epsilon) \ge 1$ *and* $(\alpha+\epsilon)v \in D$. *Then*

$$h_1(\alpha,\epsilon)\ \mathrm{conv}\{o,u\} + \alpha v \subseteq D,$$

where $h_1(\alpha,\epsilon) = \frac{\epsilon}{\sqrt{(\alpha+\epsilon)^2-1}}$.

Proof: From the hypothesis and the convexity of D, one has

$$\mathrm{conv}\{(\alpha+\epsilon)v, B\} \subseteq D. \tag{2.12}$$

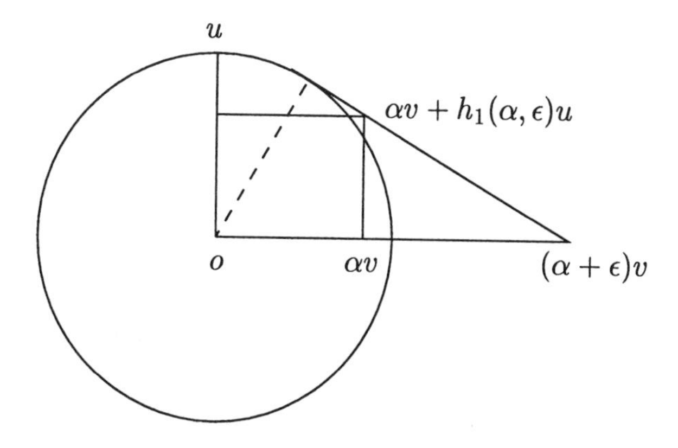

Figure 4

Also, from Figure 4 it is easy to see that

$$h_1(\alpha,\epsilon)u + \alpha v \in \operatorname{conv}\{(\alpha+\epsilon)v, B\},$$

rom which one obtains

$$h_1(\alpha,\epsilon)\operatorname{conv}\{o,u\} + \alpha v \subseteq \operatorname{conv}\{(\alpha+\epsilon)v, B\}. \tag{2.13}$$

Clearly, (2.12) and (2.13) together prove Lemma 2.7. □

Lemma 2.8. $v(G) \geq \frac{\phi}{2}$.

Proof: Let $\gamma = \langle y_1, y_2\rangle$, $\delta = \arccos(|\gamma|)$, and $\operatorname{cone}\{y_1,y_2\} = \{\alpha y_1 + \beta y_2 : \alpha \geq 0,\ \beta \geq 0\}$.

First, suppose $\gamma \geq -\frac{1}{2}$. Then we get $\delta = \phi$, $\operatorname{cone}\{y_1,y_2\} \cap B \subseteq G$, and hence

$$v(G) \geq \frac{\phi}{2}. \tag{2.14}$$

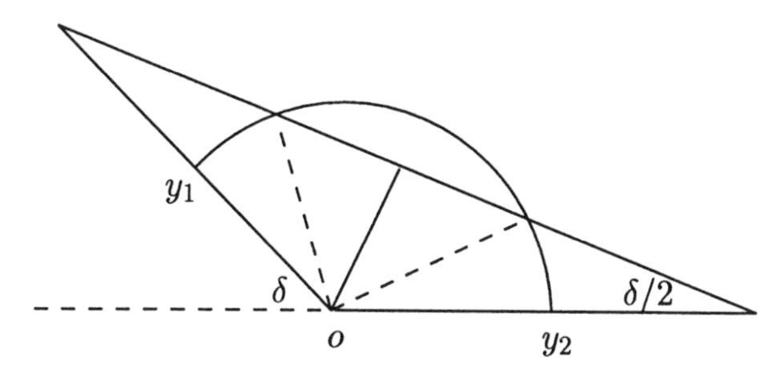

Figure 5

Next, assume $\gamma < -\frac{1}{2}$ and define $M = \text{cone}\{y_1, y_2\} \cap B \setminus G$. By som elementary calculations we get

$$v(G) = v(\text{cone}\{y_1, y_2\} \cap B) - v(M)$$
$$= \frac{\pi - \delta}{2} - \left(\arccos\left(2\sin\left(\frac{\delta}{2}\right)\right) - 2\sin\left(\frac{\delta}{2}\right)\sqrt{1 - \left(2\sin\left(\frac{\delta}{2}\right)\right)^2}\right).$$

Changing $2\sin(\frac{\delta}{2})$ to x in this expression for $v(G)$ and then using th identity $\arcsin(x) = \frac{\pi}{2} - \arccos(x)$, $0 \leq x \leq 1$, we see that

$$v(G) - \delta = \frac{\pi}{2} - \arccos(x) - \frac{3\delta}{2} + x\sqrt{1 - x^2}$$
$$= \arcsin(x) - 3\arcsin\left(\frac{x}{2}\right) + x\sqrt{1 - x^2}.$$

It follows by some routine analysis that the right-hand side is positive fc $0 \leq x \leq 1$, and so

$$v(G) - \delta \geq 0.$$

Therefore, keeping (2.10) in mind, we obtain

$$v(G) \geq \delta \geq \frac{\phi}{2}. \tag{2.15}$$

Thus, Lemma 2.8 is proven by (2.14) and (2.15). [

Before turning to the estimates on the sets $v(Q_i)$, for convenience w introduce some notation:

$$H = \{x \in R^n : \langle x, y_1 \rangle = \langle x, y_2 \rangle = 0\},$$
$$H^* = \{x = \alpha y_1 + \beta y_2 : -\infty < \alpha, \beta < +\infty\},$$
$$H_i = \{x \in R^n : \langle x, y_i \rangle = 0\}.$$

Lemma 2.9. $v(Q_1) \geq \frac{\phi}{2} h_1(1, \frac{1}{\sin(\phi)} - 1)^2 \omega_{n-2}$.

Proof: By the definition of ϕ, we have

$$|\langle y_i, y_j \rangle| \geq \cos(\phi), \quad i = 1, 2;\ j = 1,\ 2, \ldots, m - 1,$$

which implies that

$$\langle y_j, v \rangle \leq \sin(\phi), \quad v \in H_i \cap \partial(B).$$

Hence, by the definition of D, we get

$$\frac{1}{\sin(\phi)}(H_i \cap B) \subseteq D, \quad i = 1,\ 2, \tag{2.16}$$

and

$$\frac{1}{\sin(\phi)}(H \cap B) \subseteq D.$$

Applying Lemma 2.7 with $u \in H^* \cap \partial(B)$, $v \in H \cap \partial(B)$, $\alpha = 1$, and $= \frac{1}{\sin(\phi)} - 1$, we get

$$h_1\left(1, \frac{1}{\sin(\phi)} - 1\right) G + H \cap B \subseteq Q_1$$

and therefore, by applying Lemma 2.8,

$$v(Q_1) \geq \frac{\phi}{2} h_1 \left(1, \frac{1}{\sin(\phi)} - 1\right)^2 \omega_{n-2}.$$

The lemma is proven. □

Lemma 2.10. $v(Q_2) \geq h_1(1, \frac{1}{\sin(\phi)} - 1)\omega_{n-1}.$

Proof: Proceeding as in the proof of Lemma 2.9, (2.16) and Lemma 2.7 imply that

$$h_1\left(1, \frac{1}{\sin(\phi)} - 1\right) \mathrm{conv}\{o, y_i\} + H_i \cap B \subseteq Q, \quad i = 1,\ 2. \tag{2.17}$$

For convenience, denote the convex body on the left-hand side of (2.17) by J_i. Letting $z_i \in H^*$, $i = 1, 2$, be the unit vectors determined by $\langle z_i, y_i \rangle = 0$ and $\langle z_i, y_j \rangle < 0$ where $j \in \{1, 2\} \setminus \{i\}$, it is easy to see that

$$\{x \in R^n : \langle x, z_i \rangle \geq 0,\ x \in \mathrm{int}(J_i)\} \subseteq Q_2, \quad i = 1,\ 2. \tag{2.18}$$

Hence, (2.18) and the definitions of J_i and Q_2 together yield

$$v(Q_2) \geq h_1\left(1, \frac{1}{\sin(\phi)} - 1\right) \omega_{n-1},$$

which proves Lemma 2.10. □

Lemma 2.11. *If* $\langle y_1, y_2 \rangle > 0$, *then*

$$v(Q_3) \geq \frac{\pi - \phi}{2\pi} \omega_n.$$

Proof: Define

$$U = \{x \in H^* \cap B : \langle x, y_i \rangle < 0,\ i = 1, 2\}.$$

It is easy to see that

$$v(U) = \frac{\pi - \phi}{2}$$

and

$$(U + H \cap B) \cap B \subseteq Q_3.$$

Thus, we get

$$v(Q_3) \geq \frac{v(U)}{\pi}\omega_n = \frac{\pi - \phi}{2\pi}\omega_n,$$

which proves Lemma 2.11. □

Lemma 2.12. *If* $\phi \leq \frac{\pi}{4}$ *and* $\langle y_1, y_2 \rangle < 0$, *then*

$$v(Q_4) \geq \frac{\cos(\phi) - \sin(\phi)}{\cos\left(\frac{\phi}{2}\right)}\omega_{n-1}.$$

Proof: Let $u = \frac{y_1 - y_2}{\|y_1 - y_2\|}$ and $H^{**} = \{x \in R^n : \langle x, u \rangle = 0\}$. Since $\phi \leq \frac{\pi}{4} < \frac{\pi}{3}$, it is easy to see that $\langle y_j, y_1 \rangle \geq \cos(\phi)$ if and only if $\langle y_j, y_2 \rangle \leq -\cos(\phi)$, $1 \leq j \leq m-1$. It follows that

$$|\langle y_j, u \rangle| = \frac{|\langle y_j, y_1 \rangle - \langle y_j, y_2 \rangle|}{\|y_1 - y_2\|} \geq \cos(\phi),$$

which implies that

$$\langle y_j, v \rangle \leq \sin(\phi), \quad v \in H^{**} \cap B.$$

Thus, for $v \in H^{**} \cap B$ and $\lambda \in [0, 1]$, we obtain

$$\langle \lambda 2y_1 + (1-\lambda)2y_2 + v, y_j \rangle$$

$$\leq \begin{cases} \lambda(2\cos(\phi) + 2) - 2\cos(\phi) + \sin(\phi), & \langle y_j, y_1 \rangle \geq \cos(\phi) \\ -\lambda(2\cos(\phi) + 2) + 2 + \sin(\phi), & \langle y_j, y_1 \rangle \leq -\cos(\phi). \end{cases}$$

So, taking $h_2(\phi) = \frac{1 + \sin(\phi)}{2 + 2\cos(\phi)}$ and keeping the previous assumption in mind we get

$$\lambda 2y_1 + (1-\lambda)2y_2 \in B, \quad \lambda \in [h_2(\phi), 1 - h_2(\phi)],$$

$$\lambda 2y_1 + (1-\lambda)2y_2 + H^{**} \cap B \subseteq D, \quad \lambda \in [h_2(\phi), 1 - h_2(\phi)],$$

and therefore

$$\lambda 2y_1 + (1-\lambda)2y_2 + H^{**} \cap B \subseteq Q, \quad \lambda \in [h_2(\phi), 1 - h_2(\phi)]. \tag{2.19}$$

Let w be the unit vector determined by $w = \alpha y_1 + \beta y_2$ with $\alpha \geq 0, \beta \geq 0$ and $\langle w, u \rangle = 0$ and take

$$N = \{x \in R^n : x \in H^{**} \cap B, \langle w, x \rangle \geq 0\}.$$

By (2.19) and the definition of Q_4, we obtain

$$(\text{conv}\{2y_1, 2y_2\} + N) \cap Q \subseteq Q_4$$

and, hence,

$$\begin{aligned} v(Q_4) &\geq (1 - 2h_2(\phi)) \, \|2y_1 - 2y_2\| \frac{\omega_{n-1}}{2} \\ &\geq \frac{\cos(\phi) - \sin(\phi)}{\cos\left(\frac{\phi}{2}\right)} \omega_{n-1}. \end{aligned}$$

Lemma 2.12 is proven. □

Lemma 2.13. *Let $\phi > 0$ and $n > 5$. Then for every $\epsilon > 0$ such that $1 + \epsilon < \frac{2}{\sqrt{3}}$, we have*

$$v(Q_1) \geq \frac{h_1(1, \epsilon)^2 \omega_{n-2} \phi}{2(1 + h_3(1 + \epsilon, n))},$$

where

$$h_3(\mu, n) = \frac{\int_{\frac{1}{\mu}}^{1} (1 - x^2)^{\frac{n-5}{2}} \, dx}{\int_{\frac{\sqrt{3}}{2}}^{\frac{1}{\mu}} (1 - x^2)^{\frac{n-5}{2}} \, dx}, \quad 1 \leq \mu < \frac{2}{\sqrt{3}}.$$

Proof: Applying polar coordinates, it can be easily deduced that

$$v(Q_1) \geq \frac{1}{n-2} \int_G \int_{\{z \in H \cap \partial(B): \; w + z \in D\}} dz \, dw. \tag{2.20}$$

We will show that for a certain set $G^* \subseteq G$ with $v(G^*) > 0$, the above inner integral is of order ω_{n-2}. For this purpose, we set $M_\rho = \{z \in H \cap \partial(B) : \rho z \notin D\}$ and $M_\rho^* = \{z \in H \cap \partial(B) : \; \rho z \in D\}$ for $1 \leq \rho < \frac{2}{\sqrt{3}}$ and consider the inner integral at $w = o$. If $M_\rho \neq \emptyset$, then $H \cap \rho\partial(B)$ intersects the *affine hull* of certain facets F_j, say $j = 1, 2, \ldots, k$, of the Dirichlet-Voronoi cell D. Let $v_j \in H$ be the external unit normal of $\text{aff}\{F_j\} \cap H$. It follows from the special case $i = 2$ of Lemma 2.6 that

$$\text{aff}\{F_j\} \cap (H \cap \rho\partial(B)) \subseteq \text{int}(F_j \cap H), \quad 1 \leq \rho < \frac{2}{\sqrt{3}}$$

and that there exists an $\alpha_j \in [1, \rho]$ such that

$$\alpha_j v_j \in \text{int}(F_j \cap H).$$

Setting

$$M_j = \left\{ z \in H \cap \partial(B) : \; \langle z, v_j \rangle > \frac{\alpha_j}{\rho} \right\}$$

and

$$M_j^* = \left\{ z \in H \cap \partial(B) : \; \frac{\sqrt{3}\alpha_j}{2} \leq \langle z, v_j \rangle \leq \frac{\alpha_j}{\rho} \right\},$$

it is easy to see that

$$M_\rho = \bigcup_{j=1}^{k} M_j$$

and

$$\bigcup_{j=1}^{k} M_j^* \subseteq M_\rho^*.$$

In addition, setting $\gamma_z = \frac{\alpha_j}{\langle z, v_j \rangle} \geq \rho$ for $z \in M_j^*$, we get $\gamma_z z \in \text{aff}\{F_j\} \cap H$ and

$$\|\gamma_z z - \alpha_j v_j\|^2 = \langle \gamma_z z - \alpha_j v_j, \gamma_z z - \alpha_j v_j \rangle \leq \tfrac{4}{3} - \alpha_j^2.$$

This implies that $\|\gamma_z z - o\| \leq \frac{2}{\sqrt{3}}$, $\gamma_z z \in F_j \cap H$, and therefore that

$$\text{int}(M_i^*) \cap \text{int}(M_j^*) = \emptyset, \quad i \neq j.$$

Thus, we have

$$\int_{M_\rho^*} dz = \frac{\int_{M_\rho^*} dz + \int_{M_\rho} dz}{1 + \int_{M_\rho} dz / \int_{M_\rho^*} dz} \geq \frac{(n-2)\omega_{n-2}}{1 + \int_{M_j} dz / \int_{M_j^*} dz} \tag{2.21}$$

for a suitable index $j \in \{1, 2, \ldots, k\}$.

Take $g(x) = (1 - x^2)^{\frac{n-5}{2}}$ and, for $\xi \in [1, \rho)$, set

$$f_1(\xi) = \int_{\frac{\xi}{\rho}}^{1} g(x)dx \quad \text{and} \quad f_2(\xi) = \int_{\frac{\sqrt{3}\xi}{2}}^{\frac{\xi}{\rho}} g(x)dx.$$

Using polar coordinates and the monotonicity of $\frac{f_1(\xi)}{f_2(\xi)}$, we get

$$\frac{\int_{M_j} dz}{\int_{M_j^*} dz} = \frac{f_1(\alpha_j)}{f_2(\alpha_j)} \leq \frac{f_1(1)}{f_2(1)}.$$

Keeping (2.21) in mind, we obtain

$$\int_{M_\rho^*} dz \geq \frac{(n-2)\omega_{n-2}}{1 + h_3(\rho, n)}. \tag{2.22}$$

Then, applying Lemma 2.7 with $\alpha = 1$ and $\epsilon = \rho - 1$, we get

$$h_1(1, \epsilon)G + z \subseteq Q, \quad z \in M_\rho^*. \tag{2.23}$$

Hence, it follows from (2.20), (2.22), and (2.23) that for every $\epsilon > 0$ with $1 + \epsilon < \frac{2}{\sqrt{3}}$, we have

$$v(Q_1) \geq \frac{h_1(1, \epsilon)^2 \omega_{n-2} \phi}{2(1 + h_3(1 + \epsilon, n))},$$

which proves Lemma 2.13. □

Verification of the L. Fejes Tóth-Betke-Henk-Wills Phenomenon: Armed with Lemmas 2.9–2.13, the verification is an easy consequence of

$$\lim_{n\to+\infty} \frac{\omega_{n-1}}{\omega_n} = \infty.$$

We distinguish three cases depending upon ϕ and the sign of $\langle y_1, y_2\rangle$.

Case I. $0 < \phi < \frac{\pi}{4}$ *and* $\langle y_1, y_2\rangle > 0$.

By Lemmas 2.9, 2.10, and 2.11, we have that

$$\begin{aligned} v(Q) &\geq v(Q_1) + v(Q_2) + v(Q_3) \\ &\geq \omega_{n-1} + \frac{\omega_n}{2} + \left(\frac{\left(1-\sin\left(\frac{\pi}{4}\right)\right)^2 \omega_{n-2}}{2 - 2\sin^2\left(\frac{\pi}{4}\right)} - \omega_{n-1} - \frac{\omega_n}{2\pi} \right)\phi \\ &> \omega_{n-1} + \frac{\omega_n}{2} \end{aligned}$$

for all sufficiently large n.

Case II. $0 < \phi < \frac{\pi}{4}$ *and* $\langle y_1, y_2\rangle < 0$.

By Lemmas 2.9, 2.10, 2.12, and the relation $\cos\phi \geq 1 - \frac{\phi^2}{2}$, we get

$$\begin{aligned} v(Q) &\geq v(Q_1) + v(Q_2) + v(Q_4) \\ &\geq 2\omega_{n-1} + \left(\frac{\left(1-\sin\left(\frac{\pi}{4}\right)\right)^2 \omega_{n-2}}{2 - 2\sin^2\left(\frac{\pi}{4}\right)} - 2\omega_{n-1} - \frac{\pi\omega_{n-1}}{8} \right)\phi \\ &> 2\omega_{n-1} \end{aligned}$$

for all sufficiently large n.

Case III. $\phi \geq \frac{\pi}{4}$.

Choose a suitable ϵ such that the assumption of Lemma 2.13 holds. Then by Lemma 2.13 we have that

$$v(Q) \geq v(Q_1) \geq \frac{\pi h_1(1,\epsilon)^2 \omega_{n-2}}{8 + 8h_3(1+\epsilon, n)} > 2\omega_{n-1}$$

for all sufficiently large n.

Keeping Remark 2.10 in mind, it follows that case I holds for at most two Dirichlet-Voronoi cells. Thus, if n is sufficiently large and B_m is not a "sausage," we get

$$v(B_m) > 2(m-1)\omega_{n-1} + \omega_n,$$

which completes the verification of the L. Fejes Tóth-Betke-Henk-Wills phenomenon. □

Remark 2.11. *With detailed computation, it can be shown that the L. Fejes Tóth-Betke-Henk-Wills phenomenon occurs when* $n \geq 13{,}387$.

§5. Some Historical Remarks

Although it can be encountered everywhere in our daily life, finite packing is a comparatively young subject in mathematics.

Like many other geometrical problems, the earlier studies concerning finite packing were mainly confined to R^2 (see L. Fejes Tóth [3]). In particular, L. Fejes Tóth [1] showed that finite packings with B^2 cannot be denser than an optimal packing of disks in the whole plane and H. Groemer [1] found that extremal finite packings of B^2 with regard to area are essentially hexagonal parts of the densest lattice packing of disks in R^2.

Concerning higher dimensions, L. Fejes Tóth [4] raised the famous sausage conjecture in 1975 upon noticing that when $n \geq 7$, a linear arrangement of m unit balls yields a greater local density than the densest lattice packing. Since then, this conjecture has been extensively studied by U. Betke, G. Fejes Tóth, P. Gritzmann, M. Henk, J.M. Wills, and many others. Many partial results were obtained. In 1986, P. Gritzmann [1] showed that in all dimensions the sausage conjecture is true up to a constant factor by proving that for every B_m,

$$v(B_m) > \frac{2(m-1)\omega_{n-1} + \omega_n}{2+\sqrt{2}}.$$

In 1994, U. Betke, M. Henk, and J.M. Wills [1] proved the sausage conjecture for $n \geq 13,387$. Just recently, this inequality was lowered to $n \geq 45$ by M. Henk [1]. In another vein, one may ask whether the sausage conjecture remains true when one replaces the sphere by any centrally symmetric convex body. The answer to this is easily seen to be "no" when the body is a non-cylindrical *parallelohedron* since any parallelohedron tiles all of space and so has packing density one, whereas any sausage-like packing of it has local density less than one when it is noncylindrical.

In 1983, while studying the volume case of finite sphere packing in R^3 and R^4, J.M. Wills [2] found the following:

The Sausage Catastrophe: *In R^k $(k = 3, 4)$, there exists an integer m_k such that for the optimal arrangements*

$$\dim\{\text{conv}(X_m^k)\} = 1 \quad \textit{for } m < m_k,$$

while

$$\dim\{\text{conv}(X_{m_k}^k)\} = k.$$

In addition, he found that

$$m_3 \leq 56 \quad \text{and} \quad m_4 \leq 5.9 \times 10^6.$$

Besides Wills' bounds, our knowledge about m_k is very limited. A result of K. Böröczky Jr. [1] implies $m_3 \geq 4$.

J.M. Wills [1] seems to have been the first to consider other quermass-integrals. In 1991, H.T. Croft, K.J. Falconer, and R.K. Guy [1] considered finite sphere packings in general and raised the spherical conjecture. Recently, this conjecture was proven independently by K. Böröczky Jr. [2] and C. Zong [3]. The contrasting behavior of the other quermassintegrals established by Theorem 2.1 makes the L. Fejes Tóth-Betke-Henk-Wills phenomenon even stranger.

3

The Venkov-McMullen Theorem and Stein's Phenomenon

§1. Introduction

Let M be an n-dimensional compact set with interior points. If there exists a set of points X such that

$$\bigcup_{x \in X} (M + x) = R^n$$

and

$$(\mathrm{int}(M) + x_1) \cap (\mathrm{int}(M) + x_2) = \emptyset$$

whenever $x_1, x_2 \in X$ with $x_1 \neq x_2$, then we call M a *translative tile.* In addition, if X is a lattice in R^n, then we call M a *lattice tile.*

In this chapter, we will prove the famous theorem of Venkov and McMullen about convex translative tiles and convex lattice tiles. Then we will present Stein's surprising example concerning *cross translative tiles* and *cross lattice tiles.*

§2. Convex Bodies and Their Area Functions

Recall from Section 2 of Chapter 2 that every convex body K determines a unique measure $U \mapsto G(K, U)$ on $\partial(B)$ called its area function. In generalizing a theorem of H. Minkowski [2], A.D. Aleksandrov [1] and W. Fenchel and B. Jessen [1] proved the following converse, which will play a very important role in the proof of the Venkov-McMullen theorem and in the verification of Schneider's phenomenon.

Lemma 3.1. *For u any unit vector, define $B_u = \{v \in \partial(B) : \langle u, v \rangle = 0\}$. Let $G(U)$ be a positive Borel measure on $\partial(B)$ which satisfies*

$$\int_{\partial(B)} u \, G(du) = o \tag{3.1}$$

and for which

$$G(B_u) < G(\partial(B)) \tag{3.2}$$

whenever u is a unit vector. Then there exists a convex body K, unique up to translation, which has $G(U)$ as its area function.

We first recall some basics from real analysis which will be necessary for the proof of this lemma.

Definition 3.1. *A Borel set U is called a point of continuity of a positive Borel measure $G(U)$ if the values of G for U and for the interior of U are equal. A sequence of positive Borel measures $G_1, G_2, \ldots$ is said to be convergent to G if*

$$G_i(U) \to G(U)$$

for every point of continuity U of G.

Remark 3.1. *If $G_1, G_2, \ldots$ are convergent to G, then G is uniquely determined by this sequence.*

Remark 3.2. *If $K, K_1, K_2, \ldots$ are convex bodies with area functions G, $G_1, G_2, \ldots$, respectively, then $G_i \to G$ whenever $K_i \to K$.*

The verification of these remarks is routine (see H. Busemann [2]) and is thus omitted.

Proof of Lemma 3.1: We first consider the special case where there exist unit vectors u_i and positive scalars μ_i for $1 \leq i \leq N$ such that $G(\{u_i\}) = \mu_i$ and $G(U) = 0$ for $U = \partial(B) \setminus \{u_1, u_2, \ldots, u_N\}$. By (3.2), at least n of the vectors u_i are linear independent. We shall show that there exists a polytope, unique up to translation, which has N facets with external normals $u_1, u_2, \ldots, u_N$ and areas $\mu_1, \mu_2, \ldots, \mu_N$, respectively.

To show uniqueness, given $\zeta_1, \zeta_2, \ldots, \zeta_N$ non-negative numbers, define

$$P\{\zeta_i\} = \{x \in R^n : \langle x, u_j \rangle \leq \zeta_j, \ 1 \leq j \leq N\}$$

and denote the volume of $P\{\zeta_i\}$ by $v\{\zeta_i\}$. Consider the following subsets of R^{N+1}:

$$A = \{(x^1, x^2, \ldots, x^{N+1}) : x^i \geq 0, \ 1 \leq i \leq N; \ 0 \leq x^{N+1} \leq \sqrt[n]{v\{x^i\}}\}$$

and

$$A^* = \{(x^1, x^2, \ldots, x^{N+1}) :\ x^i \geq 0,\ 1 \leq i \leq N;\ x^{N+1} = \sqrt[n]{v\{x^i\}}\}.$$

From the Brunn-Minkowski inequality (Lemma 2.1), we see that

$$\sqrt[n]{v\{(1-\theta)x^i + \theta y^i\}} \geq (1-\theta)\sqrt[n]{v\{x^i\}} + \theta\sqrt[n]{v\{y^i\}}.$$

Hence, A, with A^* as part of its surface, is a cone in R^{N+1} with o as its vertex. Moreover, if $\mu_1\{\zeta_i\}$, $\mu_2\{\zeta_i\}, \ldots, \mu_N\{\zeta_i\}$ are the areas of the corresponding facets of $P\{\zeta_i\}$, the supporting hyperplane of A at $(\zeta_1, \zeta_2, \ldots, \zeta_N, \sqrt[n]{v\{\zeta_i\}})$ is

$$n\left(v\{\zeta_i\}\right)^{\frac{n-1}{n}} x^{N+1} = \sum_{j=1}^{N} \mu_j\{\zeta_i\} x^j. \tag{3.3}$$

This implies that

$$n\left(v\{\zeta_i\}\right)^{\frac{n-1}{n}} x^{N+1} \leq \sum_{j=1}^{N} \mu_j\{\zeta_i\} x^j \tag{3.4}$$

for all $(x^1, x^2, \ldots, x^{N+1}) \in A$.

Assume that there are two polytopes $P\{\zeta_i\}$ and $P\{\xi_i\}$ which have N facets with external normals u_1, $u_2, \ldots, u_N$ and corresponding areas μ_1, $\mu_2, \ldots, \mu_N$. Then, replacing $(x^1, x^2, \ldots, x^{N+1})$ in (3.4) by $(\xi_1, \xi_2, \ldots, \xi_N, \sqrt[n]{v\{\xi_i\}})$ and keeping in mind that

$$nv\{\xi_i\} = \sum_{j=1}^{N} \mu_j \xi_j,$$

it can be deduced that $v\{\xi_i\} \leq v\{\zeta_i\}$. Similarly, $v\{\zeta_i\} \leq v\{\xi_i\}$, and hence

$$v\{\zeta_i\} = v\{\xi_i\}. \tag{3.5}$$

Then it is easy to see that $(\zeta_1, \zeta_2, \ldots, \zeta_N, \sqrt[n]{v\{\zeta_i\}})$ and $(\xi_1, \xi_2, \ldots, \xi_N, \sqrt[n]{v\{\xi_i\}})$ lie in the supporting hyperplane (3.3). Therefore, the whole segment joining these two points does the same and this implies that

$$\sqrt[n]{v\{(1-\theta)\zeta_i + \theta\xi_i\}} = (1-\theta)\sqrt[n]{v\{\zeta_i\}} + \theta\sqrt[n]{v\{\xi_i\}}. \tag{3.6}$$

Since (3.6) shows that we have an instance of equality in the Brunn-Minkowski inequality, it must be the case that $P\{\zeta_i\}$ is a translate of $\lambda P\{\xi_i\}$. From (3.5) we see that $\lambda = 1$, and so

$$\zeta_i = \xi_i, \quad i = 1,\ 2, \ldots, N.$$

This proves uniqueness.

Turning now to existence, we set

$$\Phi\{\zeta_i\} = \frac{1}{n}\sum_{j=1}^{N} \mu_j \zeta_j,$$

thus moving our problem into R^N.

Obviously, $v\{\zeta_i\}$ and the $(n-1)$-dimensional volumes of the N facets of $P\{\zeta_i\}$ are continuous functions of ζ_1, $\zeta_2, \dots, \zeta_N$. Thus, the set Q of those points $(\zeta_1, \zeta_2, \dots, \zeta_N)$ which satisfy $\zeta_i \geq 0$ and $v\{\zeta_i\} \geq 1$ is closed in R^N and, hence, $\Phi\{\zeta_i\}$ attains a minimum on Q. We write this minimum in the form m^{n-1} and let $(\zeta_1^*, \zeta_2^*, \dots, \zeta_N^*)$ be a point of Q at which it is attained. It can be easily verified that $v\{\zeta_i^*\} = 1$. We now show that $mP\{\zeta_i^*\}$ is the expected polytope.

Denoting the areas of the corresponding facets of $P\{\zeta_i^*\}$ by μ_1^*, μ_2^*, ..., and μ_N^* (for convenience, we say the area of the empty set is zero), we have

$$v\{\zeta_i^*\} = \frac{1}{n}\sum_{j=1}^{N} \mu_j^* \zeta_j^* = 1 = \frac{1}{nm^{n-1}}\sum_{j=1}^{N} \mu_j \zeta_j^*. \tag{3.7}$$

Consider the following hyperplanes of R^N:

$$H_1 : \quad \frac{1}{n}\sum_{j=1}^{N} \mu_j x^j = m^{n-1}$$

and

$$H_2 : \quad \frac{1}{n}\sum_{j=1}^{N} \mu_j^* x^j = 1.$$

By (3.7), $(\zeta_1^*, \zeta_2^*, \dots, \zeta_N^*)$ belongs to both H_1 and H_2. If $x^i \geq 0$ and $(x^1, x^2, \dots, x^N)$ is a point of H_1, then we have

$$(1-\theta)(\zeta_1^*, \zeta_2^*, \dots, \zeta_N^*) + \theta(x^1, x^2, \dots, x^N) \in H_1, \quad 0 \leq \theta \leq 1,$$

and therefore

$$\Phi\{(1-\theta)\zeta_i^* + \theta x_i\} = m^{n-1}.$$

Hence, since m^{n-1} is the minimum of $\Phi\{\zeta_i\}$ on Q, we must have, on the one hand, that $v\{(1-\theta)\zeta_i^* + \theta x^i\} \leq 1$ for $0 \leq \theta \leq 1$. On the other hand,

$$\begin{aligned} v\{(1-\theta)\zeta_i^* + \theta x^i\} &= (1-\theta)^n v\left\{\zeta_i^* + \frac{\theta}{1-\theta}x^i\right\} \\ &\geq (1-\theta)^n \left(v\{\zeta_i^*\} + \frac{n\theta}{1-\theta}V_1(P\{\zeta_i^*\}, P\{x^i\})\right) \\ &\geq 1 - n\theta + n(1-n\theta)\theta V_1(P\{\zeta_i^*\}, P\{x^i\}). \end{aligned}$$

Thus, $V_1(P\{\zeta_i^*\}, P\{x^i\}) \leq 1$, and so by (2.2),

$$\frac{1}{n}\sum_{j=1}^{N} \mu_j^* x^j \leq 1.$$

This implies that H_1 lies in one side of H_2. Recalling that $(\zeta_1^*, \zeta_2^*, \ldots, \zeta_N^*) \in H_1 \cap H_2$, we obtain $H_1 = H_2$, which shows that $mP\{\zeta_i^*\}$ is the expected polytope.

Now we deal with the general case by approximation.

Decomposing $\partial(B)$ into a finite number of Borel sets with diameters less than $\frac{1}{l}$, denote by U_1, $U_2, \ldots, U_N$ those pieces of the decomposition to which G assigns positive mass, and then define unit vectors u_i and positive scalars γ_i by

$$\int_{U_i} uG(du) = G(U_i)\gamma_i u_i, \quad 1 \leq i \leq N.$$

Finally, define a new positive Borel measure $G_l(U)$ by

$$G_l(U) = \sum_{u_i \in U} G(U_i)\gamma_i.$$

Then $G_l(U)$ satisfies (3.1). Given any continuous function $g(u)$ on $\partial(B)$, let $g_l(u)$ be the function which has the value $g(u_i)\gamma_i$ on each U_i for $i = 1, 2, \ldots, N$, and the value $g(u)$ on $\partial(B) \setminus \bigcup_{j=1}^{N} U_j$. Then

$$\int_{\partial(B)} g(u)G_l(du) = \int_{\partial(B)} g_l(u)G(du).$$

Since

$$\cos\left(\frac{1}{l}\right) \leq \gamma_i \leq 1,$$

it is clear that γ_i tends uniformly to 1 as $l \to +\infty$, and so $g_l(u)$ tends uniformly to $g(u)$. Therefore,

$$\lim_{l\to\infty} \int_{\partial(B)} g(u)G_l(du) = \int_{\partial(B)} g(u)G(du).$$

Since $g(u)$ was any continuous function, it follows that

$$\lim_{l\to+\infty} G_l(U) = G(U). \tag{3.8}$$

For an arbitrary $u \in \partial(B)$, define

$$T_u(v) = \max\{0, \langle u, v\rangle\}.$$

Obviously, $T_u(v)$ is positive on the open *hemisphere* $B_u^* = \{x \in \partial(B) : \langle x, u\rangle > 0\}$ and vanishes on the rest of $\partial(B)$. Hence, keeping (3.2) in mind, it can be shown that there exists a positive $c > 0$ such that for all $u \in \partial(B)$,

$$\int_{\partial(B)} T_u(v)G(dv) \geq 2c.$$

Also, since $\{T_u(v) : u \in \partial(B)\}$ is trivially equicontinuous,

$$\lim_{l\to+\infty} \int_{\partial(B)} T_u(v)G_l(dv) = \int_{\partial(B)} T_u(v)G(dv)$$

uniformly in u. This implies that there exists an l_0 such that for all $l \geq l_0$ and all $u \in \partial(B)$,

$$\int_{\partial(B)} T_u(v)G_l(dv) > c.$$

From this, it follows that $G_l(U)$ satisfies (3.2) for all $l \geq l_0$. In what follows, we work only with the measures $G_{l_0}, G_{l_0+1}, \dots$.

Applying the special case already proven to each G_l, we get a polytope sequence $P_{l_0}, P_{l_0+1}, \dots$ such that each G_l is the area function of P_l. By (3.8), we see that there exists a number k such that for all $l \geq l_0$,

$$s(P_l) = G_l(\partial(B)) \leq k.$$

Then, by applying (2.3) with $L = B$, we get that for all $l \geq l_0$,

$$v(P_l) \leq \left(\frac{k^n}{n^n\omega_n}\right)^{\frac{1}{n-1}}.$$

If the segment $[o, \beta u]$ lies in P_l, then the supporting funtion of P_l is at least $\beta T_u(v)$. Therefore, keeping Remark 2.1 in mind, we get

$$nv(P_l) \geq \beta \int_{\partial(B)} T_u(v)G_l(dv) \geq c\beta,$$

which implies that

$$P_l \subseteq c^{-1}nv(P_l)B \subseteq c^{-1}n\left(\frac{k^n}{n^n\omega_n}\right)^{\frac{1}{n-1}} B.$$

Hence, applying Blaschke's selection theorem, we obtain a subsequence P_{l_1}, $P_{l_2}, \dots$ of $P_{l_0}, P_{l_0+1}, \dots$ which converges to a convex body K. Then by (3.8), Remark 3.1, and Remark 3.2, we get that K is a convex body which has $G(U)$ as its area function.

Since uniqueness in the general case will not be needed in this book, we omit its proof. □

With some minor modifications in the previous proof, one can prove the following.

Corollary 3.1. *Let $N > n$. For fixed non-negative numbers $\mu_1, \mu_2, \dots, \mu_N$ and distinct unit vectors $u_1, u_2, \dots, u_N$ which contain n independent ones, there exists a unique centrally symmetric convex polytope which has N pairs of facets with normals $u_1, -u_1; u_2, -u_2; \dots; u_N, -u_N$, and areas $\mu_1, \mu_1; \mu_2, \mu_2; \dots; \mu_N, \mu_N$, respectively.*

§3. The Venkov-McMullen Theorem

Theorem 3.1 (B.A. Venkov [1], P. McMullen [2]). *Every convex translative tile is also a lattice tile.*

As preparation for the proof of this theorem, let us introduce some notation and definitions. For P an n-dimensional convex polytope, let F_1^i, F_2^i, $\ldots, F_{k(P,i)}^i$ denote the i-dimensional faces of P. When all the facets F_1^{n-1}, $F_2^{n-1}, \ldots, F_{k(P,n-1)}^{n-1}$ as well as the polytope P itself are centrally symmetric, we call the collection $\mathcal{F}_i$ of those facets which contain a translate of a given $(n-2)$-dimensional face F_i^{n-2} a *belt* of P.

Lemma 3.2 (H. Minkowski [2]). *If a polytope P can be decomposed into finitely many nonoverlapping centrally symmetric convex polytopes, then P itself is centrally symmetric.*

Proof: Assume that P is the union of m nonoverlapping centrally symmetric polytopes P_1, $P_2, \ldots, P_m$. Denote by $F(u)$ and $F_i(u)$ the facets of P and P_i, respectively, which take u as their external normals, and by $f(u)$ and $f_i(u)$ their areas. Since the P_i are centrally symmetric, it is easy to see that $f_i(u) = f_i(-u)$ for every unit vector u. It follows from the hypothesis that we can find a subset A_1 of $\{1, 2, \ldots, m\}$ such that

$$F(u) = \bigcup_{i \in A_1} F_i(u). \tag{3.9}$$

Obviously, if $F_i(-u) \cap F(-u) \neq \emptyset$, then $F_i(-u) \subseteq F(-u)$. Hence, the set A_1 can be divided into two disjoint subsets

$$A_1^1 = \{i \in A_1 : \; F_i(-u) \subseteq F(-u)\}$$

and

$$A_1^2 = \{i \in A_1 : \; F_i(-u) \cap F(-u) = \emptyset\}.$$

This and (3.9) imply that

$$f(u) \leq \sum_{i \in A_1^1} f_i(-u) + \sum_{i \in A_1^2} f_i(-u). \tag{3.10}$$

We now consider A_1^2. Since P is convex and is the union of P_1, P_2, $\ldots$, P_m, there exists a subset A_2 of $\{1, 2, \ldots, m\} \setminus A_1$ such that

$$\bigcup_{i \in A_1^2} F_i(-u) \subseteq \bigcup_{i \in A_2} F_i(u).$$

Dividing A_2 into

$$A_2^1 = \{i \in A_2 : \; F_i(-u) \subseteq F(-u)\}$$

and

$$A_2^2 = \{i \in A_2 : F_i(-u) \cap F(-u) = \emptyset\},$$

we obtain

$$\sum_{i \in A_1^2} f_i(-u) \le \sum_{i \in A_2^1} f_i(-u) + \sum_{i \in A_2^2} f_i(-u). \tag{3.11}$$

Continuing in this manner, it is clear that this process can be repeated at most m times. Thus, by (3.10) and (3.11), we get

$$\begin{aligned} f(u) &\le \sum_{i \in A_1^1} f_i(-u) + \sum_{i \in A_2^1} f_i(-u) + \cdots \\ &\le f(-u). \end{aligned}$$

In a similar fashion, one obtains $f(-u) \le f(u)$, and therefore $f(u) = f(-u)$ for every direction u.

The lemma then follows immediately from Corollary 3.1. □

Lemma 3.3. *Every convex translative tile P is a centrally symmetric polytope with centrally symmetric facets such that each belt of P contains four or six facets.*

Proof: Assume that X is a set of points which contains o such that

$$\bigcup_{x \in X} (P + x) = R^n \tag{3.12}$$

and

$$\text{int}(P + x_1) \cap \text{int}(P + x_2) = \emptyset$$

whenever $x_1, x_2 \in X$ with $x_1 \ne x_2$.

First, note that it is well known, and also easy to see, that for two distinct points x and y of X, $P + x$ meets $P + y$ if and only if $x - y \in \partial(D(P))$. Hence, $\|x - y\|$ is bounded and bounded away from zero. Consequently, there are only finitely many translates $P + t_i$ ($i = 1, 2, \ldots, l$) of $P + X$ which touch P. Applying a basic result of Convex and Discrete Geometry, we see that there are l hyperplanes which separate P and $P + t_i$ respectively for $i = 1, 2, \ldots, l$. Then, keeping (3.12) in mind, it follows that P is a convex polytope.

Second, by (3.12) it is obvious that

$$\bigcup_{x \in X} (-P - x) = R^n$$

and that the corresponding translates are nonoverlapping. Hence, there are finitely many points, say $x_1, x_2, \ldots, x_m \in X$, such that

$$P \subseteq \bigcup_{i=1}^{m} (-P - x_i),$$

and so

$$P = \bigcup_{i=1}^{m} ((-P - x_i) \cap P).$$

Since $(-P - x_i) \cap P$, $i = 1, 2, \ldots, m$, are centrally symmetric and nonoverlapping, Lemma 3.2 implies that P must be centrally symmetric.

Third, using the notation introduced in the proof of Lemma 3.2, we can easily see that there are finitely many points $y_1, y_2, \ldots, y_p \in X$ such that

$$F(u) \subseteq \bigcup_{i=1}^{p} (F(-u) + y_i)$$

and, hence,

$$F(u) = \bigcup_{i=1}^{p} \Big((F(-u) + y_i) \cap F(u) \Big).$$

Obviously, $\big(F(-u)+y_i\big) \cap F(u)$, $i = 1, 2, \ldots, \ldots, p$, are centrally symmetric and nonoverlapping. Hence, applying Lemma 3.2 once more, we get that every facet $F(u)$ of P is centrally symmetric.

Finally, letting the belt of P determined by an $(n-2)$-dimensional face F_i^{n-2} consist of q pairs of opposite facets $F_1^{n-1}, \ldots, F_{2q}^{n-1}$, we must prove that $q \leq 3$. By considering $(n-2)$-dimensional *Lebesgue measure*, it is easy to see that there is a point $z \in F_i^{n-2}$ which lies on no j-dimensional face of any translate $P + x$ with $j < n - 2$ and $x \in X$. Denote those translates which contain z by $P + z_i$ $(i = 1, 2, \ldots, w)$. Let Γ be the projection from R^n to the 2-dimensional plane which contains z and is perpendicular to F_i^{n-2}. It is easy to see that $\Gamma(P)$ is a centrally symmetric polygon which has z as a vertex and $\Gamma(F_1^{n-1})$, $\ldots, \Gamma(F_{2q}^{n-1})$ as its edges. Moreover, $\Gamma(P + z_1)$, $\ldots, \Gamma(P + z_w)$ are nonoverlapping translates of $\Gamma(P)$ which have z in their boundary. Recall from elementary geometry that the sum of the interior angles of $\Gamma(P)$ is $2(q-1)\pi$. Thus, the sum of g $(g < q)$ mutually nonopposite angles of $\Gamma(P)$ is greater than

$$\frac{2(q-1)\pi - (2q - 2g)\pi}{2} = (g-1)\pi.$$

From this, one can see that z is a vertex of each of the polygons $\Gamma(P + z_i)$.

Since $P + z_1, \ldots, P + z_w$, and P join properly at z, from consideration of $\Gamma(P + z_1), \ldots, \Gamma(P + z_w)$, and $\Gamma(P)$ at z it can be seen that

$$\frac{2(q-1)\pi}{2} \leq 2\pi.$$

This implies that $q \leq 3$, and so Lemma 3.3 is proven. □

Letting P be a convex translative tile, by Lemma 3.3 all the facets of P and P itself are centrally symmetric. Assume that o is the center of P and that F_1^{n-1}, $-F_1^{n-1}, \dots, F_m^{n-1}$ and $-F_m^{n-1}$ are the m pairs of facets of P. Define vectors f_i for $i = 1, 2, \dots, m$ by the condition

$$-F_i^{n-1} + f_i = F_i^{n-1}$$

and then set

$$\Lambda = \left\{ \sum_{i=1}^{m} n_i f_i : \; n_i \in \mathbb{Z} \right\}.$$

We shall prove Theorem 3.1 by showing that

$$P + \Lambda = R^n \tag{3.13}$$

and that

$$\big(\mathrm{int}(P) + x_1\big) \cap \big(\mathrm{int}(P) + x_2\big) = \emptyset \tag{3.14}$$

whenever $x_1, x_2 \in \Lambda$ with $x_1 \neq x_2$.

Assume that F is a face of a translate $P + x$ where $x \in \Lambda$. Denote by $X(x, F)$ the subset of Λ defined recursively by (i) $x \in X(x, F)$ and (ii) $z \in X(x, F)$ whenever there exists a $y \in X(x, F)$ such that $P + y$ and $P + z$ meet in a common facet which contains F. Since Λ is a lattice, we can restrict our attention to $X(o, F)$ when studying the local properties of $X(x, F)$.

Lemma 3.4. *Let* $\mathrm{int}(F)$ *be the relative interior of* F. *Then*

$$\mathrm{int}(F) \subset \mathrm{int}\left(\bigcup_{x \in X(o,F)} (P + x) \right).$$

Proof: Denote the dimension of F by $\dim(F)$. Clearly, the lemma holds if $\dim(F) = n - 1$. Supposing that the lemma is true when $\dim(F) \geq r + 1$, we must prove that it is also true when $\dim(F) = r$. To this end, let F be an r-dimensional face of P, z a relative interior point of F, and

$$B^* = \partial(\epsilon B^{n-r}) + z$$

a small $(n - r)$-dimensional sphere which is orthogonal to F and which meets only those facets of $P + x$, $x \in X(o, F)$, which contain F. It suffices to prove that

$$B^* \subset \bigcup_{x \in X(o,F)} (P + x). \tag{3.15}$$

For convenience, we write

$$\begin{aligned} Q(x, F) &= \bigcup_{\substack{F \subset F' \\ \dim(F') > r}} \; \bigcup_{y \in X(x,F')} (P + y) \\ &= \bigcup_{\substack{F \subset F' \\ \dim(F') = r+1}} \; \bigcup_{y \in X(x,F')} (P + y). \end{aligned}$$

It is easy to see that each point of $(P+x)\cap B^*$, $x\in X(o,F)$, lies in the relative interior of some face F' of $P+x$ with $\dim(F')>r$. By our inductive assumption, each point of $B^*\cap \operatorname{int}(F')$ has a *neighborhood* in B^* contained in

$$\bigcup_{y\in X(x,F')}(P+y)\subseteq Q(x,F).$$

Since $(P+x)\cap B^*$ is compact, it follows that there exists a $\delta>0$ such that the δ-neighborhood of $(P+x)\cap B^*$ is contained in $Q(x,F)$. As $X(o,F)$ is a finite set, we can use the same δ for each $x\in X(o,F)$. Hence, the δ-neighborhood in B^* of each point of

$$B'=B^*\cap\left(\bigcup_{x\in X(o,F)}(P+x)\right)$$

is contained in

$$\bigcup_{x\in X(o,F)}Q(x,F)=\bigcup_{x\in X(o,F)}(P+x).$$

Since B^* is connected and B' is compact, open, and nonempty, (3.15) is proven, and so Lemma 3.4 follows immediately. □

Lemma 3.5. *If x_1 and x_2 are two distinct points of $X(o,F)$, then*

$$\operatorname{int}(P+x_1)\cap\operatorname{int}(P+x_2)=\emptyset.$$

Proof: When $\dim(F)=n-1$, the assertion of the lemma is obvious. When $\dim(F)=n-2$, repeating the observation at the end of the proof of Lemma 3.3, it is easy to see that the assertion of the lemma is true. So, assuming that the assertion of the lemma is true for all faces F with $\dim(F)\geq r+1$, it only remains to show the same for an arbitrary r-dimensional face F.

Suppose, to the contrary, that there exists a face F with $\dim(F)=r$ and two distinct points $x_1, x_2\in X(o,F)$ such that $\operatorname{int}(P+x_1)\cap\operatorname{int}(P+x_2)\neq\emptyset$. Then by the definition of $X(o,F)$, we can find l points $y_1=x_1, y_2,\ldots,y_l=x$ of $X(o,F)$ such that each $(P+y_i)\cap(P+y_{i+1})$, containing F, is a common facet of $P+y_i$ and $P+y_{i+1}$. In this situation, we call the sequence $P+y_1,\ P+y_2,\ldots,P+y_l$ an *$\{x_1,x_2\}$-chain.* If $z\in\operatorname{int}(F)$ and B^* is the small sphere introduced in Lemma 3.4, then this $\{x_1,x_2\}$-chain gives rise to a corresponding $\{x_1,x_2\}^*$-chain of *spherical* $(n-r-1)$*-polytopes* $P_i^*=(P+y_i)\cap B^*$ for $i=1,2,\ldots,l$.

Choosing $p\in\operatorname{int}(P_1^*\cap P_l^*)$, we consider the *spherical polygonal loops* L based at p; that is, we consider those L composed of a sequence, beginning and ending at p, of finitely many arcs of great circles of B^*. Such a loop L is said to be associated with an $\{x_1,x_2\}^*$-chain $P_1^*,\ P_2^*,\ldots,P_l^*$ if L passes successively from P_{i-1}^* to P_i^* for $i=2,\ 3,\ldots,l$. In this situation, L is a union of l subarcs $L_i\subset P_i^*$, some of which possibly degenerate to points, such that L_{i-1} and L_i meet at endpoints of each other in $P_{i-1}^* \cap P_i^*$. Of course, L_l and L_1 meet at p. Moreover, we call L an *interior loop* if each L_i is contained in

$$\mathrm{int}(P_i^*) \cup \mathrm{int}(P_{i-1}^* \cap P_i^*) \cup \mathrm{int}(P_i^* \cap P_{i+1}^*).$$

Denote the length of a spherical loop L by $\lambda(L)$ and denote the infimum of $\lambda(L)$ over all interior loops based at p and associated with some $\{x_1, x_2\}^*$-chain by λ. We are going to prove that $\lambda = 0$, which implies that $P_1^* = P_l^*$, and so $P + x_1 = P + x_2$.

In doing this, we will use the fact that the dimension of B^* is at least 2, so that B^* is simply connected with the consequence that loops in B^* based at p can be contracted within B^* to p.

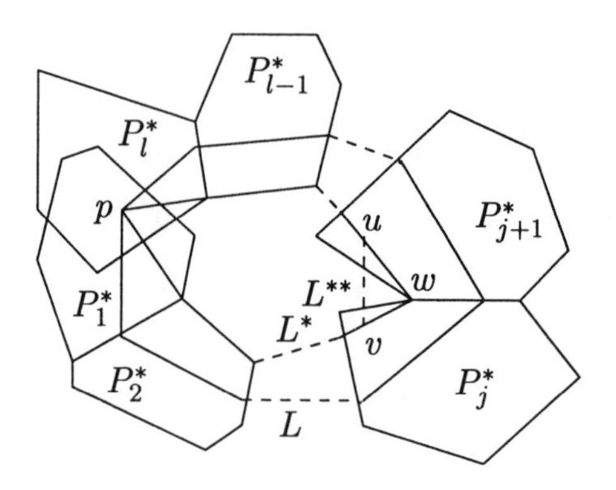

Figure 6

If $\lambda \neq 0$, then a topological argument involving the Hausdorff metric shows that there exists a loop L^* associated with an $\{x_1, x_2\}^*$-chain $P_1^* = (P + x_1) \cap B^*$, $P_2^*, \ldots, P_l^* = (P + x_2) \cap B^*$ such that $\lambda(L^*) = \lambda$. Clearly, L^* is not an interior loop. Hence, L^* has a nonstraight angle at the relative interior of a face $F^* \cap B^*$ of some P_j^*, where F^* is a face of the corresponding $P + y_j$ with $y_j \in X(o, F)$, $F \subset F^*$, and

$$\dim(F^*) \geq r + 1. \tag{3.16}$$

Assuming that L^* meets F^* at w, it follows from Lemma 3.4 that there exists a δ-neighborhood $N(w)$ of w in

$$B^* \cap \left(\bigcup_{y \in X(y_j, F^*)} (P + y) \right).$$

Choose $u, v \in L^* \cap N(w)$ such that w is between u and v on $L^* \cap N(w)$. Then obtain a new loop L^{**} by replacing that part of L^* between u and v

by that arc $\widehat{uv}$ of of the great circle of B^* determined by u and v for which

$$\widehat{uv} \subset N(w) \subset \operatorname{int}\left(B^* \cap \bigcup_{y \in X(y_j, F^*)} (P+y)\right). \tag{3.17}$$

Clearly, $\lambda(L^{**}) < \lambda$, and so a contradiction will be obtained if we show that L^{**} is associated with some $\{x_1, x_2\}^*$-chain.

Suppose s is the smallest index such that $u \in P_s^*$ and t is the largest index such that $v \in P_t^*$. It is easy to see that $s \neq t$. The finiteness of $X(y_j, F^*)$ guarantees that we can find two sequences $u_k \subset N(w) \cap \operatorname{int}(P_s^*)$ with $\lim_{k\to+\infty} u_k = u$ and $v_k \subset N(w) \cap \operatorname{int}(P_t^*)$ with $\lim_{k\to+\infty} v_k = v$ such that $\widehat{u_k v_k}$ meets no face of dimension less than $n-1$ of any $P+y$, $y \in X(y_j, F^*)$, and such that $\widehat{u_k v_k}$ passes through the same sequence $Q_1^* = P_s^*$, $Q_2^*, \ldots, Q_h^* = P_t^*$, where $Q_i^* = (P + y_i^*) \cap B^*$, $y_i^* \in X(y_j, F^*)$. Then by (3.16), (3.17), and the inductive assumption, we see that $(P+y_i^*) \cap (P+y_{i+1}^*)$ is a common facet of both $P+y_i^*$ and $P+y_{i+1}^*$ containing F and F^*. Finally, since

$$\lim_{k\to+\infty} \widehat{u_k v_k} = \widehat{uv},$$

we see that we obtain a new $\{x_1, x_2\}^*$-chain with which L^{**} is associated when we replace $P_{s+1}^*, \ldots, P_{t-1}^*$ in the original $\{x_1, x_2\}^*$-chain by Q_2^*, $\ldots, Q_{h-1}^*$.

Obviously, now $\lambda = 0$, and so $x_1 = x_2$, thus completing the proof of the lemma. □

Proof of Theorem 3.1: As has already been noted, it suffices to verify (3.13) and (3.14).

If (3.13) does not hold, we can find a maximal connected set D such that

$$D \subseteq R^n \setminus \bigcup_{x \in \Lambda} (P+x). \tag{3.18}$$

Since $\partial(D)$ contains a point which belongs to the relative interior of a facet of some translate $P+x$, $x \in \Lambda$, (3.18) contradicts the definition of Λ. Thus, (3.13) is true.

If (3.14) does not hold, we can find two distinct points x_1, $x_2 \in \Lambda$ such that $\operatorname{int}(P+x_1) \cap \operatorname{int}(P+x_2) \neq \emptyset$. As in the proof of Lemma 3.5, we take $p \in \operatorname{int}(P+x_1) \cap \operatorname{int}(P+x_2)$ and denote by L a loop containing p. We call $P+y_1$, $P+y_2, \ldots, P+y_l$ an $\{x_1, x_2\}$-chain when $y_1 = x_1$, $y_l = x_2$, and $(P+y_i) \cap (P+y_{i+1})$ is a facet of both $P+y_i$ and $P+y_{i+1}$ for each i. In addition, say that L is associated with this chain if L begins at p, passes successively from $P+y_i$ to $P+y_{i+1}$, and then ends at p. Finally, let λ be the infimum of $\lambda(L)$, the length of L, over all those loops L which are associated with some $\{x_1, x_2\}$-chain. As in the proof of Lemma 3.5, a loop associated with some $\{x_1, x_2\}$-chain exists whose length is λ. Lemma 3.4

allows us to find a suitable shorter loop, and then Lemma 3.5 ensures us that this loop is associated with some $\{x_1, x_2\}$-chain, a contradiction which establishes (3.14).

Our proof of Theorem 3.1 is now complete. □

§4. Stein's Phenomenon

In some sense, *Stein's phenomenon* can be thought of as a counterpart to the Venkov-McMullen Theorem. To present Stein's phenomemon, which is an example, some definitions and notation are first necessary.

Cross and Notched Cross: *Let ϵ and δ be small positive numbers. Denote by $I(m,n)$ the set of points on the coordinate axes of R^n whose coordinates are integers with modulus $\leq m$, by $Q'(n,\epsilon)$ the set of points which have only two nonzero coordinates, these being $|x^i| = \frac{1}{2} + \epsilon$ and $x^{i+1} = \frac{\text{sign}\{x^i\}}{4}$ (consider $x^{n+1} = x^1$), and by $Q^*(n,\epsilon)$ the set of points which also have only two non-zero coordinates, these now being $|x^i| = \frac{1}{2} - \epsilon$ and $x^{i+1} = -\frac{\text{sign}\{x^i\}}{4}$ (again consider $x^{n+1} = x^1$). Recalling that W is the unit cube in R^n centered at the origin, set*

$$W^* = W \cup \big((1+\delta)\epsilon B + Q^*(n,\epsilon)\big) \setminus \big((1+\delta)\,\epsilon B + Q'(n,\epsilon)\big).$$

Then we call

$$M(m,n) = I(m,n) + W$$

and

$$M(m,n,\epsilon,\delta) = I(m,n) + W^*$$

an (m,n)-cross and a notched (m,n)-cross, respectively. It is obvious that both $M(m,n)$ and $M(m,n,\epsilon,\delta)$ are centrally symmetric star bodies. In addition, $M(m,n,\epsilon,\delta)$ can be obtained by digging $2n(2mn-2m+1)$ suitable holes into and adding $2n(2mn-2m+1)$ suitable caps onto the surface of $M(m,n)$.

With these concepts, S.K. Stein [2] in 1972 found

Stein's Phenomenon: *$M(4,10,\epsilon,\delta)$ tiles R^{10}, but not as a lattice.*

To verify Stein's discovery, we need two basic results, one from Number Theory and the other from Abstract Algebra.

Let Y be a set with q elements and consider $Y^n = Y \oplus Y \oplus \cdots \oplus Y$, the n-fold *Cartesian product* of Y with itself. We call a subset U of Y^n a *perfect cover* of Y^n when each element of Y^n differs from exactly one member of U in at most one coordinate.

Lemma 3.6 (S.K. Zaremba [1]). *Let q be a power of a prime and l, greater than or equal to 2, be an integer. Set $n = \frac{q^l-1}{q-1}$. Then there exists a perfect cover of Y^n consisting of q^{n-l} elements.*

Proof: Assume that Y is a *finite field* of q elements. Then Y^n is a vector space of dimension n over Y. Letting V denote a vector space of dimension l over Y, the number of 1-dimensional subspaces of V is

$$\frac{q^l - 1}{q - 1} = n.$$

Choosing n nonzero vectors $v_1, v_2, \ldots, v_n$ of V in turn from its n 1-dimensional subspaces and letting $e_1, e_2, \ldots, e_n$ be the standard basis of Y^n, we define a linear transformation T from Y^n onto V by

$$T(e_i) = v_i, \quad i = 1, 2, \ldots, n.$$

Let U be the *kernel* of T. Then, since T is onto, U has dimension $n-l$ over Y, and so

$$\text{card}\{U\} = q^{n-l}. \tag{3.19}$$

Given $x \in Y^n$, $T(x) = av_i$ for some $a \in Y$ and some v_i. Clearly, then

$$T(x - ae_i) = av_i - av_i = o,$$

and therefore

$$x - ae_i = w \in U,$$

i.e., x differs from $w \in U$ in at most the i-th coordinate. The uniqueness of w is easy to check. Hence, U is a perfect cover of Y^n.

Keeping (3.19) in mind, we see that the proof of Lemma 3.6 is complete. □

Let G be a group and let A_1 and A_2 be subsets of G. If each element of G can be uniquely expressed in the form $a_1 a_2$ with $a_1 \in A_1$ and $a_2 \in A_2$, then G is called a product of A_1 and A_2. We need the following result involving group *homomorphisms* and this notion of product:

Lemma 3.7. *Assume that a group G^* is a product of subsets A_1^* and A_2^*. Let G be another group and $f\colon G \mapsto G^*$ be a homomorphism from G onto G^*. Let A_1 be any subset of G such that the restriction of f to A_1 is a bijection between A_1 and A_1^*, and set $A_2 = f^{-1}(A_2^*)$. Then G is a product of A_1 and A_2.*

Proof: First, we show that every element $x \in G$ is of the required form. By hypothesis, $f(x) = a_1^* a_2^*$ for some $a_1^* \in A_1^*$ and $a_2^* \in A_2^*$. Taking $a_1 \in A_1$ such that $f(a_1) = a_1^*$, we see that

$$f(a_1^{-1}x) = (a_1^*)^{-1} a_1^* a_2^* = a_2^*.$$

Hence, $a_2 = a_1^{-1}x$ is an element in A_2. Thus, $x = a_1a_2$ with $a_1 \in A_1$ and $a_2 \in A_2$.

Next, we establish uniqueness. Assume that $a_1a_2 = b_1b_2$, where $a_1, b_1 \in A_1$ and $a_2, b_2 \in A_2$. Then we have

$$f(a_1)f(a_2) = f(b_1)f(b_2).$$

Since G^* is a product of A_1^* and A_2^*, we get that $f(a_1) = f(b_1)$. Consequently, $a_1 = b_1$, and then $a_2 = b_2$, proving Lemma 3.7. □

Verification of Stein's Phenomenon: Taking $q = 3^2$ and $n = \frac{q^2-1}{q-1} = 10$ in Lemma 3.6, we see that for any set Y with nine elements, the Cartesian product Y^{10} has a perfect cover U. Choose Y to be $\mathbb{Z}_9 = \mathbb{Z}/9\mathbb{Z}$, the *cyclic group* of order nine. As is well known, $Y^{10} = \mathbb{Z}_9^{10}$ forms a group, $\mathbb{Z}_9 \oplus \mathbb{Z}_9 \oplus \cdots \oplus \mathbb{Z}_9$, when addition is defined in the obvious componentwise manner. For $z \in \mathbb{Z}$, let $\overline{z}$ denote the element $z + 9\mathbb{Z}$ of $\mathbb{Z}_9$. Taking

$$A_1^* = \bigcup_{i=1}^{10}\{(\overline{z^1},\overline{z^2},\ldots,,\overline{z^{10}}) \in \mathbb{Z}_9^{10} : \ \overline{z^i} = \overline{-4},\ldots,\overline{4};\ \overline{z^j} = \overline{0},\ j \neq i\},$$

it is easy to see that $\operatorname{card}\{A_1^*\} = 81$. Let $A_2^* = U$. Since U is a perfect cover of $\mathbb{Z}_9^{10}$, $\mathbb{Z}_9^{10}$ is a product of A_1^* and A_2^*.

Let f be the homomorphism $f : \mathbb{Z}^{10} \mapsto \mathbb{Z}_9^{10}$ defined by

$$f(z^1, z^2, \ldots, z^{10}) = (\overline{z^1}, \overline{z^2}, \ldots, \overline{z^{10}})$$

and take $A_1 = I(4, 10) \subset \mathbb{Z}^{10}$. Clearly, $f : A_1 \mapsto A_1^*$ is a bijection between A_1 and A_1^*; so by Lemma 3.7, $\mathbb{Z}^{10}$ is a product of A_1 and $f^{-1}(A_2^*)$. Thus $M(4, 10)$, and hence also $M(4, 10, \epsilon, \delta)$, tiles R^{10} with $X = f^{-1}(A_2^*)$.

Next, we show that $M(4, 10, \epsilon, \delta)$ is not a lattice tile. Assume on the contrary that $M(4, 10, \epsilon, \delta)$ is a lattice tile associated with a lattice Λ. By considering the special structure of $M(4, 10, \epsilon, \delta)$, it is easy to see that $\Lambda \subset \mathbb{Z}^{10}$. Thus, $\mathbb{Z}^{10}$ would be the product of the set A_1, consisting of the 81 elements described above, and a subgroup H of $\mathbb{Z}^{10}$. Let

$$h : \mathbb{Z}^{10} \to \mathbb{Z}^{10}/H$$

be the natural homomorphism. Observe that $\mathbb{Z}^{10}/H$ has order 81 and that

$$h : A_1 \to \mathbb{Z}^{10}/H$$

is a bijection. Let $g_i \in \mathbb{Z}^{10}/H$ be $h(e_i)$, where the e_i, $i = 1, 2, \ldots, 10$, are the standard unit vectors of $\mathbb{Z}^{10}$. Thus, the 80 nonzero elements of $\mathbb{Z}^{10}/H$ are $\pm g_i$, $\pm 2g_i$, $\pm 3g_i$, and $\pm 4g_i$, $i = 1, 2, \ldots, 10$. Clearly, the order of each g_i is at least 9.

Now the group $\mathbb{Z}^{10}/H$ is either $\mathbb{Z}_3 \oplus \mathbb{Z}_3 \oplus \mathbb{Z}_3 \oplus \mathbb{Z}_3$, $\mathbb{Z}_9 \oplus \mathbb{Z}_9$, $\mathbb{Z}_{27} \oplus \mathbb{Z}_3$, $\mathbb{Z}_9 \oplus \mathbb{Z}_3 \oplus \mathbb{Z}_3$, or $\mathbb{Z}_{81}$. The case $\mathbb{Z}_3 \oplus \mathbb{Z}_3 \oplus \mathbb{Z}_3 \oplus \mathbb{Z}_3$ is ruled out since all its elements have order less than 9. The case $\mathbb{Z}_9 \oplus \mathbb{Z}_9$ is ruled out since each g_i would have order 9, and then the 20 elements $\pm 3g_i$ would have order 3. But there are only eight elements of order 3 in $\mathbb{Z}_9 \oplus \mathbb{Z}_9$. To rule out $\mathbb{Z}_{27} \oplus \mathbb{Z}_3$, we note that $\mathbb{Z}_{27} \oplus \mathbb{Z}_3$ has eight elements of order 3. Thus, four of the elements $g_1, g_2, \ldots, g_{10}$, say g_1, g_2, g_3, and g_4, would have order 9, and $\pm 3g_1$, $\pm 3g_2$, $\pm 3g_3$, and $\pm 3g_4$ would be the eight elements of order 3. Then the 24 elements $\pm g_i$, $\pm 2g_i$, $\pm 4g_i$, $i = 1, 2, 3, 4$, would be of order 9. However, $\mathbb{Z}_{27} \oplus \mathbb{Z}_3$ has only 18 elements of order 9. To rule out $\mathbb{Z}_9 \oplus \mathbb{Z}_3 \oplus \mathbb{Z}_3$, note that, on the one hand, $\mathbb{Z}_9 \oplus \mathbb{Z}_3 \oplus \mathbb{Z}_3$ has 26 elements of order 3, but, on the other hand, only the 20 elements $\pm 3g_1, \pm 3g_2, \ldots, \pm 3g_{10}$ of $\mathbb{Z}^{10}/H$ may be of order 3. Finally, the case $\mathbb{Z}_{81}$ can be ruled out as follows. Consider $\mathbb{Z}_{81}$ to be the set $\{0, 1, 2, \ldots, 80\}$ under addition modulo 81. Let $J = \{i \in \mathbb{Z}_{81} : 1 \leq i \leq 80, (i, 81) = 1\}$. Then J contains $\varphi(3^4) = 54$ elements. Without loss of generality, $g_1 = 1$. If

$$\mathbb{Z}_{81} \setminus \{o\} = \bigcup_{i=1}^{10} \{\pm g_i, \pm 2g_i, \pm 3g_i, \pm 4g_i\}, \tag{3.20}$$

then J, regarded as a group under multiplication modulo 81, may be assumed to be the product of $\{g_1, g_2, \ldots, g_9\}$ and $\{\pm 1, \pm 2, \pm 4\}$. It is then easy to show that the set $\{g_1, g_2, \ldots, g_9\}$ is, up to changes of sign, $\{8^0, 8^1, \ldots, 8^8\}$. But $3 \cdot 8^0 \equiv 3 \bmod (81)$, whereas $(-3) \cdot 8^3 \equiv 3 \bmod (81)$. Thus, $3 \cdot 8^0 \equiv (-3) \cdot 8^3 \bmod (81)$ and hence the decomposition (3.20) cannot exist.

In conclusion, we see that $M(4, 10, \epsilon, \delta)$ cannot be a lattice tile, and so the verification of Stein's phenomenon is complete. □

§5. Some Remarks

As one of the most natural and most ancient of mathematical subjects, tiling has been studied by many famous mathematicians. These include Aleksandrov, Aristotle, Delone, Dirichlet, Euclid, Fedorov, Minkowski, Venkov, Voronoi, and many living mathematicians. Besides being an important and fruitful subject in pure mathematics, tiling is also practical and finds application in crystallography and many other sciences. For some general surveys of tiling, we refer the reader to P. Engel [2], P. Erdös, P.M. Gruber, and J. Hammer [1], B. Grünbaum and G.C. Shephard [2], and E. Schulte [1].

Within the general realm of tiling problems, the more restricted problems connected with translative tiles and lattice tiles have attracted great interest among modern geometers. In 1885, E.S. Fedorov [1] determined the two "types" of convex lattice tiles in R^2 and the five "types" of convex lattice tiles in R^3. This work gave rise to a fascinating, but very difficult, body of research. Because of the deep and exhaustive investigations of B.N. Delone [1], M.I. Štogrin [1], and P. Engel [1], today we know that there are

52 "types" of 4-dimensional convex lattice tiles. As is the case with many other geometric problems, we know almost nothing about answers, in this case the classification of the convex lattice tiles, in yet higher dimensions.

Another direction of research is embodied in Theorem 3.1, which was first proved by B.A. Venkov [1] in 1954. Almost 30 years later, this deep result was independently rediscovered by P. McMullen [2]. Clearly, Theorem 3.1 is counterintuitive and somewhat hard to believe. A.D. Aleksandrov [3] and H. Groemer [4] have obtained some generalizations and refinements of this result.

Going further, it is very natural to ask whether Theorem 3.1 remains true when we omit the convexity condition on the tile. In 1972, S.K. Stein [2] answered this question negatively with his example, and the result has come to be known as Stein's phenomenon. In 1981, S. Szabó [1] managed to reduce the dimension from 10 to 5 in Stein's phenomenon. Recently, Szabó has also constructed examples in spaces of dimension $n = 2^k - 1$ for $k \geq 2$, thereby settling the problem in R^3 (see S. Szabó [3]).

To end this chapter, for contrast's sake we mention the works of L. Fejes Tóth, C.A. Rogers, and M.R. von Wolff concerning packing densities of 2-dimensional domains. In 1950, L. Fejes Tóth [2] and C.A. Rogers [1] proved that for every convex domain, the density of its densest translative packing and the density of its densest lattice packing are identical. In 1962, M.R. von Wolff [1] found that this result of Fejes Tóth and Rogers does not hold for star domains.

4

Local Packing Phenomena

1. Introduction

et K be a fixed convex body in R^n. We call the largest number of nonover-
pping translates of K which can be brought into contact with K the *kiss-
ng number* of K and denote it by $h(K)$. A closely related but contrasting
oncept is the *blocking number* of K, denoted $z(K)$, which is the smallest
umber of nonoverlapping translates of K which are in contact with K
nd prevent any other translate of K from touching K. Concerning kiss-
ng numbers and blocking numbers, one can raise the following intuitive
roblem:

roblem 4.1. *Let K_1 and K_2 be two distinct convex bodies in R^n. Does
$(K_1) < h(K_2)$ always imply that $z(K_1) \leq z(K_2)$?*

Let Λ be a lattice in R^n with *determinant* $d(\Lambda)$. If $K + u_1$ and $K + u_2$
re nonoverlapping for any distinct points u_1 and u_2 of Λ, we call $K + \Lambda$ a
ttice packing of K and Λ a *packing lattice* of K. The quantity $\delta(K, \Lambda) =$
$(K)/d(\Lambda)$ is called the density of this packing, whereas the number $h(K, \Lambda)$
f translates $K + u_i$ which are in contact with K and for which $u_i \in \Lambda \setminus \{o\}$
called the Λ-*kissing number* of K. Denoting the family of all packing
ttices of K by $\mathcal{L}(K)$, we call

$$\delta^*(K) = \sup_{\Lambda \in \mathcal{L}(K)} \{\delta(K, \Lambda)\}$$

and

$$h^*(K) = \max_{\Lambda \in \mathcal{L}(K)} \{h(K, \Lambda)\}$$

the *lattice packing density* and the *lattice kissing number* of K, respectivel
Finally, we define

$$\mathcal{L}^\delta(K) = \{\Lambda \in \mathcal{L}(K) : \ \delta(K, \Lambda) = \delta^*(K)\}$$

and

$$\mathcal{L}^h(K) = \{\Lambda \in \mathcal{L}(K) : \ h(K, \Lambda) = h^*(K)\} .$$

As usual, the lattices belonging to $\mathcal{L}^\delta(K)$ are called the *densest packin lattices* of K. By *Mahler's selection theorem*, it is easy to see that bot $\mathcal{L}^\delta(K)$ and $\mathcal{L}^h(K)$ are nonempty and even compact in the sense of the nat ural topology[1] on lattices. With these definitions, another natural probler immediately arises:

Problem 4.2. *Consider two distinct lattices Λ_1 and Λ_2 in $\mathcal{L}(K)$. Doe $\Lambda_1 \in \mathcal{L}^\delta(K)$ and $h(K, \Lambda_1) < h(K, \Lambda_2)$ always imply that $\Lambda_2 \in \mathcal{L}^\delta(K)$?*

Intuition suggests that both Problems 4.1 and 4.2 have affirmative an swers. For the first problem, one might argue that the more nonoverlap ping translates one can bring into contact with a fixed translate, the mor nonoverlapping translates one needs to prevent other translates from touch ing the fixed translate. For the second problem, one might argue that every translate of K from a lattice packing $K + \Lambda$ is in contact with mor translates of K from $K + \Lambda$ than in another lattice packing, then a larg box should contain more translates of K from $K + \Lambda$ than from this othe lattice packing, and hence that the packing density of $K + \Lambda$ should be th larger. Moreover, in R^2 both Problems 4.1 and 4.2 do indeed have affirma tive answers. Nevertheless, as we shall see in this chapter, such intuitio and low-dimensional evidence can lead one astray.

---

[1]This topology is defined by specifiying the ϵ-neighborhood of a given lattic $A\mathbb{Z}^n$ to be

$$\left\{ \Lambda = B\mathbb{Z}^n : \ d(\Lambda) \neq 0, \ \left(\sum_{i,j} (a_{ij} - b_{ij})^2 \right)^{\frac{1}{2}} < \epsilon \right\} .$$

§2. A Phenomenon Concerning Blocking Numbers and Kissing Numbers

In attacking Problem 4.1, C. Zong [4] found the following.

Phenomenon α: *When n is sufficiently large, we can find two convex bodies K_1 and K_2 in R^n such that both*

$$h(K_1) < h(K_2) \quad \textit{and} \quad z(K_1) > z(K_2)$$

hold simultaneously.

Lemma 4.1 (R.A. Rankin [1]). *Let $x_i = (x_i^1, x_i^2, \ldots, x_i^n)$, $i = 1, 2, \ldots, m$, be m distinct points on the boundary of the n-dimensional unit ball B, and define*

$$d = \min_{\substack{1\le i,j\le m\\ i\ne j}} \{\|x_i - x_j\|\}.$$

Then for any point $x \in \partial(B)$, we have

$$\left(\sum_{i=1}^{m} d_i^2\right)^2 - 4m\sum_{i=1}^{m} d_i^2 + 2m(m-1)d^2 \le 0,$$

where $d_i = \|x - x_i\|$.

Proof: We may suppose, without loss of generality, that $x = (1, 0, \ldots, 0)$. Then, from the way d is defined above, we have that

$$\begin{aligned}
\frac{1}{2}m(m-1)d^2 &\le \sum_{k=1}^{n} \sum_{1\le i<j\le m} \left(x_i^k - x_j^k\right)^2 \\
&= \sum_{k=1}^{n}\left\{ m\sum_{i=1}^{m}\left(x_i^k\right)^2 - \left(\sum_{i=1}^{m} x_i^k\right)^2\right\} \\
&= m\sum_{i=1}^{m}\left(1 - x_i^1\right)^2 - \left(m - \sum_{i=1}^{m} x_i^1\right)^2 \\
&\quad + \sum_{k=2}^{n}\left\{ m\sum_{i=1}^{m}\left(x_i^k\right)^2 - \left(\sum_{i=1}^{m} x_i^k\right)^2\right\} \\
&= m\sum_{i=1}^{m}\left\{\left(1 - x_i^1\right)^2 + \sum_{k=2}^{n}\left(x_i^k\right)^2\right\} \\
&\quad - \left\{\sum_{i=1}^{m}\left(1 - x_i^1\right)\right\}^2 - \sum_{k=2}^{n}\left(\sum_{i=1}^{m} x_i^k\right)^2.
\end{aligned}$$

Since

$$d_i^2 = \left(1 - x_i^1\right)^2 + \sum_{k=2}^{n} \left(x_i^k\right)^2 = 2\left(1 - x_i^1\right),$$

we deduce that

$$\frac{1}{2}m(m-1)d^2 \le m\sum_{i=1}^{m} d_i^2 - \frac{1}{4}\left(\sum_{i=1}^{m} d_i^2\right)^2 - \sum_{k=2}^{n}\left(\sum_{i=1}^{m} x_i^k\right)^2,$$

which implies the lemma. □

Lemma 4.2 (R.A. Rankin [1], C. Zong [4]). *For the n-dimensional unit ball B, we have*

$$h(B) \ll \pi^{\frac{1}{2}} n^{\frac{3}{2}} 2^{\frac{n-1}{2}} \tag{4.1}$$

and

$$z(B) \ge \frac{2\pi^{\frac{1}{2}}}{n}\left(\frac{4}{3}\right)^{\frac{n-3}{2}}, \tag{4.2}$$

where $f(n) \ll g(n)$ means $f(n) \le g(n)$ when n is large.

Proof: Let ρ be the spherical distance on $\partial(B)$. First, we show (4.1).

Let $X = \{o, 2x_1, 2x_2, \ldots, 2x_{h(B)}\}$ be a set of points such that $B + X$ is a finite packing of B and such that $B \cap (B + 2x_i) \neq \emptyset$ for $i = 1, 2, \ldots, h(B)$. Clearly,

$$x_i \in \partial(B) \quad \text{for } i = 1, 2, \ldots, h(B)$$

and

$$\rho(x_i, x_j) \ge \frac{\pi}{3} \quad \text{for } i \neq j.$$

Hence, the caps $G_i = \{x \in \partial(B) : \rho(x, x_i) \le \frac{\pi}{6}\}$, $i = 1, 2, \ldots, h(B)$, form a *cap packing* of $\partial(B)$.

Replace each G_i by a concentric cap $G_i^* = \{x \in \partial(B) : \rho(x, x_i) \le \frac{\pi}{4}\}$ and attach a density

$$\delta_i(x) = 2\sqrt{2}\left(\cos\theta - \frac{\sqrt{2}}{2}\right)$$

to points x of G_i^* at a spherical distance θ from x_i. Note that one can also describe $\delta_i(x)$ as $\sqrt{2}\left(\gamma^2 - r^2\right)$, where $\gamma = 2\sin\frac{\pi}{8}$ is the Euclidean distance from the boundary of G_i^* to x_i and $r = 2\sin\frac{\theta}{2}$ is the Euclidean distance from x to x_i. Then the total "mass" of G_i^* is

$$\begin{aligned} M &= \int_0^{\frac{\pi}{4}} (n-1)\omega_{n-1}(\sin\theta)^{n-2}\delta_i(x)d\theta \\ &= 2\sqrt{2}(n-1)\omega_{n-1}\int_0^{\frac{\pi}{4}} (\sin\theta)^{n-2}\left(\cos\theta - \frac{\sqrt{2}}{2}\right)d\theta. \end{aligned} \tag{4.3}$$

Let z be any point of $\partial(B)$. Then either z belongs to no cap G_i^*, in which case the total density at z is zero, or z belongs to $m \geq 1$ such caps centered at $x_1, x_2, \ldots, x_m$ say, in which case the total density at z is

$$\delta(z) = \sum_{i=1}^{m} \delta_i(z) = \sqrt{2}\left(m\gamma^2 - \sum_{i=1}^{m} d_i^2\right),$$

where $d_i = \|z - x_i\|$.

Hence,

$$\begin{aligned}\sum_{i=1}^{m} d_i^2 &= m\gamma^2 - \frac{\sqrt{2}}{2}\delta(z)\\ &= 4\left(\sin\frac{\pi}{8}\right)^2\left(m - \frac{1+\sqrt{2}}{2}\delta(z)\right),\end{aligned}$$

and so by Lemma 4.1,

$$\begin{aligned}0 \geq\ & 16\left(\sin\frac{\pi}{8}\right)^4\left(m - \frac{1+\sqrt{2}}{2}\delta(z)\right)^2\\ & -16m\left(\sin\frac{\pi}{8}\right)^2\left(m - \frac{1+\sqrt{2}}{2}\delta(z)\right) + 2m(m-1).\end{aligned}$$

This reduces to

$$4m(1-\delta(z)) \geq \delta^2(z),$$

which implies that $\delta(z) \leq 1$. Hence, we see that $h(B)M$ is less than the total surface area of $\partial(B)$, and so by (4.3),

$$\begin{aligned}h(B) &\leq \frac{n\omega_n}{M}\\ &= \frac{\sqrt{2}n\omega_n}{4(n-1)\omega_{n-1}\int_0^{\frac{\pi}{4}}(\sin\theta)^{n-2}\left(\cos\theta - \frac{\sqrt{2}}{2}\right)d\theta}.\end{aligned} \tag{4.4}$$

We now estimate the integral in (4.4) from below. Setting

$$t = (n-1)\log\frac{1}{\sqrt{2}\sin\theta},$$

we get

$$\begin{aligned}I_1 &= \frac{\sqrt{2}}{2}\int_0^{\frac{\pi}{4}}(\sin\theta)^{n-2}d\theta\\ &= \frac{1}{n-1}2^{\frac{1-n}{2}}\int_0^{\infty}\frac{e^{-t}dt}{\left\{1+\left(1-e^{\frac{-2t}{n-1}}\right)\right\}^{\frac{1}{2}}}.\end{aligned}$$

Using Lagrange's form of the remainder in *Taylor's theorem*, we easily see that for every $x \geq 0$,

$$(1+x)^{-\frac{1}{2}} \leq 1 - \frac{1}{2}x + \frac{3}{8}x^2,$$

and so for $n \geq 3$,

$$\begin{aligned} I_1 &\leq \frac{1}{n-1} 2^{\frac{1-n}{2}} \int_0^\infty e^{-t} \left\{ 1 - \frac{1}{2}\left(1 - e^{\frac{-2t}{n-1}}\right) + \frac{3}{8}\left(1 - e^{\frac{-2t}{n-1}}\right)^2 \right\} dt \\ &= \frac{1}{n-1} 2^{\frac{1-n}{2}} \left\{ 1 - \frac{n}{(n+1)(n+3)} \right\} \\ &\leq \frac{1}{n-1} 2^{\frac{1-n}{2}} - \frac{1}{2n^2} 2^{\frac{1-n}{2}}. \end{aligned}$$

Combining this inequality with the fact that

$$I_2 = \int_0^{\frac{\pi}{4}} (\sin\theta)^{n-2} \cos\theta \, d\theta = \frac{1}{n-1} 2^{\frac{1-n}{2}},$$

we see that for $n \geq 3$,

$$\int_0^{\frac{\pi}{4}} (\sin\theta)^{n-2} \left(\cos\theta - \frac{\sqrt{2}}{2} \right) d\theta = I_2 - I_1 \geq \frac{1}{2n^2} 2^{\frac{1-n}{2}}. \tag{4.5}$$

Since

$$\frac{\omega_n}{\omega_{n-1}} = \frac{\Gamma(1 + \frac{n-1}{2})}{\Gamma(1 + \frac{n}{2})} \pi^{\frac{1}{2}} \ll \frac{\sqrt{2}\pi^{\frac{1}{2}}}{n^{\frac{1}{2}}},$$

obviously (4.4) and (4.5) together yield (4.1).

We now turn to (4.2). Let $Y = \{2y_1, 2y_2, \ldots, 2y_{z(B)}\}$ be such that $B + \{Y \cup \{o\}\}$ is a finite packing, $B \cap (B + 2y_i) \neq \emptyset$ holds for $i = 1, 2, \ldots, z(B)$, and $\cup_{i=1}^{z(B)} (B + 2y_i)$ prevents any other translate of B from touching B. Clearly, $y_i \in \partial(B)$ for $i = 1, 2, \ldots, z(B)$. Defining $Q_i = \{x \in \partial(B) : \rho(x, y_i) \leq \frac{\pi}{3}\}$, it is easy to verify that

$$\bigcup_{i=1}^{z(B)} Q_i = \partial(B). \tag{4.6}$$

In other words, the caps $Q_1, Q_2, \ldots, Q_{z(B)}$ form a *cap covering* of $\partial(B)$.

Letting A_i be the surface area of Q_i and keeping (4.6) in mind, we note that

$$\sum_{i=1}^{z(B)} A_i \geq n\omega_n$$

and that

$$A_i = (n-1)\omega_{n-1} \int_{\frac{1}{2}}^{1} (1-x^2)^{\frac{n-3}{2}} dx$$
$$\leq (n-1)\omega_{n-1} \left(\frac{1}{2}\right) \left(\frac{3}{4}\right)^{\frac{n-3}{2}}.$$

Hence,

$$z(B) \geq \frac{n\omega_n}{(n-1)\omega_{n-1} \left(\frac{1}{2}\right) \left(\frac{3}{4}\right)^{\frac{n-3}{2}}}$$
$$= \frac{2n\pi^{\frac{1}{2}}}{n-1} \frac{\Gamma(1+\frac{n-1}{2})}{\Gamma(1+\frac{n}{2})} \left(\frac{4}{3}\right)^{\frac{n-3}{2}}$$
$$\geq \frac{2\pi^{\frac{1}{2}}}{n} \left(\frac{4}{3}\right)^{\frac{n-3}{2}},$$

thus finishing the proof of Lemma 4.2. □

Lemma 4.3 (C. Zong [4]). *Letting ϵ be a positive number, define*

$$T_{i,\epsilon}(x^1, \ldots, x^i, \ldots, x^n) = (1-\epsilon|x^i|) \left(x^1, \ldots, \frac{x^i}{1-\epsilon|x^i|}, \ldots, x^n \right)$$

and

$$T = T_{1,\epsilon} T_{2,\epsilon} \cdots T_{n,\epsilon}.$$

Then, setting $Q = T(W)$, where W is the n-dimensional unit cube centered at the origin o, we have, for ϵ sufficiently small, that

$$h(Q) \geq 2^n \quad \textit{and} \quad z(Q) \leq 2n.$$

Proof: For the course of this proof, rather than use the usual metric on R^n, it is more convenient to employ the "maximum" metric

$$d^*(x,y) = \max_{1 \leq i \leq n} \{|x^i - y^i|\}$$

to measure the distance between two points x and y.

Clearly, Q is a centrally symmetric convex body contained in W. By choosing $\epsilon > 0$ sufficiently small, one may make the d^*-distance between x and $T(x)$ less than $1/4$ for any point $x \in W$. In consequence, for every pair of distinct vertices x and y of W,

$$d^*(2T(x), 2T(y)) \geq 1.$$

Therefore, if $x_1, x_2, \ldots, x_{2^n}$ are the 2^n vertices of W, all the translates $Q + 2T(x_1), \ldots, Q + 2T(x_{2^n})$ are pairwise nonoverlapping and in contact with Q in a nonoverlapping manner. This means that

$$h(Q) \geq 2^n. \tag{4.7}$$

Setting $y_1 = (1, 0, \ldots, 0)$, $y_2 = (0, 1, 0, \ldots, 0)$, $\ldots$, $y_n = (0, \ldots, 0, 1)$, it can be easily verified that all the translates $Q \pm y_1, \ldots, Q \pm y_n$ are pairwise nonoverlapping and in contact with Q in a nonoverlapping manner. On the one hand, by the construction of Q we notice that

$$\left\{(x^1, x^2, \ldots, x^n) \in \partial(2Q) : \; x^i > \tfrac{1}{2}\right\} \subset \operatorname{int}(2Q) + y_i.$$

On the other hand, since the d^*-distance between x and $T(x)$ is less than $1/4$ for any point $x \in W$, every boundary point z of $2Q$ has a coordinate z^i such that

$$|z^i| > \tfrac{1}{2}.$$

Hence, we see that

$$\partial(2Q) \subset \bigcup_{i=1}^{n} \left(\operatorname{int}(2Q) \pm y_i\right),$$

which implies that $\cup_{i=1}^{n}(Q \pm y_i)$ prevents any other translate of Q from touching Q. Thus, we have

$$z(Q) \leq 2n. \tag{4.8}$$

Because of (4.7) and (4.8), Lemma 4.3 is proven. □

Obviously, Lemma 4.2 and Lemma 4.3 together show that Phenomenon α holds with $K_1 = B$ and $K_2 = Q$ when n is sufficiently large.

§3. A Basic Approximation Result

In this section we state and prove a fundamental approximation result which will be frequently needed in the rest of this book.

Definition 4.1. *A convex body K is called regular if every boundary point of K lies on exactly one supporting hyperplane of K and every supporting hyperplane of K meets exactly one boundary point of K.*

With this definition out of the way, we may proceed.

Lemma 4.4. *For any fixed convex body K and any given positive number η, there is a regular convex body K^* such that*

$$\delta^H(K^*, K) < \eta.$$

Proof: Suppose that $K \subseteq r'B$. Given $r > r'$, let $I(r)$ denote the intersection of all the balls of radius r which contain K.

We first prove that $\lim_{r \to +\infty} \delta^H(I(r), K) = 0$.

If $\delta^H(I(r), K) \not\to 0$, since $I(r_2) \subseteq I(r_1)$ whenever $r_1 < r_2$, there would be a point x belonging to every $I(r)$ for sufficiently large r and at a positive distance μ from K. Letting y be the nearest point of K to such an x, it would then follow from some elementary calculations that both

$$K \subseteq y + \gamma \frac{y-x}{\mu} + \sqrt{d^2(K) + \gamma^2} B$$

and

$$x \notin y + \gamma \frac{y-x}{\mu} + \sqrt{d^2(K) + \gamma^2} B$$

for sufficiently large γ. This contradiction proves that

$$\lim_{r \to +\infty} \delta^H(I(r), K) = 0.$$

By what has just been proved, we may choose r so large that $\delta^H(I(r), K) < \frac{1}{2}\eta$. Setting $K^* = I(r) + \frac{1}{2}\eta B$, it is easy to see that $K \subseteq K^*$ and $\delta^H(K^*, K) < \eta$. To finish, it suffices to show that K^* is regular.

On the one hand, if $p \in \partial(K^*)$, then there is a point $z \in I(r)$ such that $\|p - z\| = \frac{1}{2}\eta$. Hence, $z + \frac{1}{2}\eta B \subset K^*$; thus there is a unique supporting hyperplane of K^* at p. On the other hand, if a supporting hyperplane H of K^* meets $\partial(K^*)$ at two distinct points p_1 and p_2, then H will meet $\partial(K^*)$ at the closed segment p_1p_2. This implies that for u the external normal of H,

$$p_1p_2 - \tfrac{1}{2}\eta u \subset \partial(I(r)),$$

a contradiction.

Thus, Lemma 4.4 is proven. □

Upon restricting ourselves to the family of centrally symmetric convex bodies, we have the following:

Corollary 4.1. *For any fixed centrally symmetric convex body C and any given positive number η, there is a regular centrally symmetric convex body C^* such that*

$$\delta^H(C^*, C) < \eta.$$

§4. Minkowski's Criteria for Packing Lattices and the Densest Packing Lattices

In order to evaluate the density of the densest lattice packings of a given convex body K in R^3, it is necessary in this section to discuss *Minkowski's criteria* for a lattice to be a packing lattice of K and a densest packing lattice of K. However, some preliminaries must be attended to beforehand.

Lemma 4.5. *Let $a_1, a_2, \ldots, a_n$ be n independent points of a lattice Λ. Then Λ has a basis $\{u_1, u_2, \ldots, u_n\}$ in which $a_1, a_2, \ldots, a_n$ have the following* *representation:*

$$a_i = \sum_{j=1}^{i} \mu_{i,j} u_j,$$

where $0 \leq \mu_{i,j} < \mu_{i,i}$ *whenever* $j < i$.

Proof: Let $L_i = L(a_1, a_2, \ldots, a_i)$ be the subspace generated by $a_1, a_2, \ldots, a_i$ and take $v_i \in \Lambda \cap L_i$ such that

$$d(v_i, L_{i-1}) = \min_{v \in \Lambda \cap L_i \setminus L_{i-1}} \{d(v, L_{i-1})\}.$$

It is easy to see that $\{v_1, v_2, \ldots, v_n\}$ is a basis of Λ and that each a_i is a linear combination of the vectors $v_1, v_2, \ldots, v_i$ only.

Inductively, if $u_1, u_2, \ldots, u_{i-1}$ have been chosen and

$$a_i = \nu_{i,i} v_i + \sum_{j=1}^{i-1} \nu_{i,j} u_j,$$

then we may take

$$u_i = \pm v_i - \xi_{i,1} u_1 - \cdots - \xi_{i,i-1} u_{i-1},$$

where the sign and the integers $\xi_{i,j}$ remain to be chosen. Clearly, $\{u_1, u_2, \ldots, u_n\}$ is a basis of Λ and

$$a_i = (\nu_{i,1} - \nu_{i,i}\xi_{i,1})u_1 + \cdots + (\nu_{i,i-1} - \nu_{i,i}\xi_{i,i-1})u_{i-1} \pm \nu_{i,i} u_i.$$

Hence, we may choose the sign so that

$$\mu_{i,i} = \pm \nu_{i,i} > 0$$

and the integers $\xi_{i,j}$ so that

$$0 \leq \mu_{i,j} = \nu_{i,j} - \nu_{i,i}\xi_{i,j} < \mu_{i,i}, \quad j = 1, 2, \ldots, i-1.$$

Thus, Lemma 4.5 is proven. □

Corollary 4.2. *Let* a_1 *and* a_2 *be two independent points of a lattice* Λ *in* R^2. *Suppose that the triangle with vertices* o, a_1, *and* a_2, *i.e., the convex hull of* $\{o, a_1, a_2\}$, *contains no other points of* Λ. *Then* $\{a_1, a_2\}$ *is a basis of* Λ.

Lemma 4.6. *Let* x_1 *and* x_2 *be two distinct points and let* K *be a convex set. Then* $(K + x_1) \cap (K + x_2) \neq \emptyset$ *if and only if*

$$\left(\tfrac{1}{2}D(K) + x_1\right) \cap \left(\tfrac{1}{2}D(K) + x_2\right) \neq \emptyset.$$

Since the proof of this lemma is routine (see P.M. Gruber and C.G. Lekkerkerker [1]), we omit it. By Lemma 4.6, we have that

$$\mathcal{L}(K) = \mathcal{L}\left(\tfrac{1}{2}D(K)\right)$$

and

$$\mathcal{L}^{\delta}(K) = \mathcal{L}^{\delta}\left(\tfrac{1}{2}D(K)\right).$$

Hence, since $\frac{1}{2}D(K)$, being centrally symmetric, is easier to handle than K, in what follows we will, in effect, deal with $\mathcal{L}(\frac{1}{2}D(K))$ and $\mathcal{L}^{\delta}(\frac{1}{2}D(K))$ instead of $\mathcal{L}(K)$ and $\mathcal{L}^{\delta}(K)$.

The following classic result of Swinnerton-Dyer is a necessary preliminary to Minkowski's criteria.

Lemma 4.7 (H.P.F. Swinnerton-Dyer [1]). *Let C be a centrally symmetric convex body in R^n. Then for every $\Lambda \in \mathcal{L}^{\delta}(C)$,*

$$h(C, \Lambda) \geq n(n+1).$$

Proof: Without loss of generality, $\Lambda = \mathbb{Z}^n$. Set

$$\Lambda' = (I + \eta A)\mathbb{Z}^n,$$

where I is the unit matrix, $A = (a_{ij})$ is a matrix to be chosen below, and $|\eta|$ is small. Let $\pm p_1, \ldots, \pm p_m$ be the points Λ such that $C \cap (p_i + C) \neq \emptyset$. Supposing that $m < \frac{1}{2}n(n+1)$, we will show that there exist suitable A and η such that $\Lambda' \in \mathcal{L}(C)$ and $d(\Lambda') < 1 = d(\Lambda)$, which contradicts the membership of Λ in $\mathcal{L}^{\delta}(C)$.

Let u_k be an external unit normal of $2C$ at p_k and set $q_k = (I + \eta A)p_k$ for $k = 1, 2, \ldots, m$. We can arrange it so that each q_k belongs to the supporting hyperplane

$$H_k = \{x \in R^n : \langle x - p_k, u_k \rangle = 0\}$$

by choosing the elements a_{ij} of the matrix A so as to satisfy the equations

$$\langle Ap_k, u_k \rangle = \sum_{i,j} a_{ij} p_k^j u_k^i = 0 \quad \text{for } k = 1, 2, \ldots, m. \tag{4.9}$$

It is even possible to choose the a_{ij}, not all zero, so that the m relations of (4.9) hold and, so that $a_{ij} = a_{ji}$ for all $i \neq j$. This is because we have $m + \frac{1}{2}n(n-1) < n^2$. Now observe that the points q_k do not belong to the interior of $2C$. It is then clear that $\Lambda' \in \mathcal{L}(C)$ if $|\eta|$ is sufficiently small.

Turning to $d(\Lambda')$, we have

$$d(\Lambda') = d(I + \eta A) = 1 + \beta_1 \eta + \beta_2 \eta^2 + \cdots + \beta_n \eta^n,$$

where

$$\beta_1 = \sum_{i=1}^{n} a_{ii}, \quad \beta_2 = \sum_{i<j} (a_{ii}a_{jj} - a_{ij}a_{ji}), \ \ldots .$$

Note that

$$2\beta_2 - \beta_1^2 = -2\sum_{i<j} a_{ij}a_{ji} - \sum_{i=1}^{n} a_{ii}^2$$

$$= \frac{1}{2}\sum_{i<j}\{(a_{ij} - a_{ji})^2 - (a_{ij} + a_{ji})^2\} - \sum_{i=1}^{n} a_{ii}^2.$$

Since $a_{ij} = a_{ji}$ for $i \neq j$ and not all of the a_{ij} are zero, we conclude that $2\beta_2 - \beta_1^2 < 0$. Hence, either (i) $\beta_1 \neq 0$ or (ii) $\beta_1 = 0$ and $\beta_2 < 0$. In either case, one may arrange for $d(\Lambda') < 1$ by choosing η to have small enough magnitude and appropriate sign.

Thus, $h(C, \Lambda) = 2m \geq n(n+1)$, and so Lemma 4.7 is proven. □

Let Λ be a lattice in R^3 and let O be an octahedron with center at o. If o and the six vertices of O are the only points of $O \cap \Lambda$, then we call O a *lattice octahedron* of Λ.

Lemma 4.8. *Let O be a lattice octahedron of Λ with vertices $\pm v_i$ ($i = 1, 2, 3$). Then Λ has a basis $\{u_1, u_2, u_3\}$ satisfying one of the two following alternatives*:

I. $v_i = u_i$ *for* $i = 1, 2, 3$;

II. $v_i = u_i$ *for* $i = 1, 2$ *and* $v_3 = u_1 + u_2 + 2u_3$.

Proof: According to Lemma 4.5, Λ has a basis $\{u_1, u_2, u_3\}$ in which the three vertices v_1, v_2, and v_3 of O have the form

$$v_1 = \mu_{1,1}u_1, \quad v_2 = \mu_{2,1}u_1 + \mu_{2,2}u_2, \quad v_3 = \mu_{3,1}u_1 + \mu_{3,2}u_2 + \mu_{3,3}u_3,$$

where $0 < \mu_{1,1}$, $0 \leq \mu_{2,1} < \mu_{2,2}$, $0 \leq \mu_{3,1} < \mu_{3,3}$, and $0 \leq \mu_{3,2} < \mu_{3,3}$. Since O is a lattice octahedron, the triangle ov_1v_2 contains no other points of Λ. Hence, by Corollary 4.2 we have that $\mu_{1,1} = \mu_{2,2} = 1$ and $\mu_{2,1} = 0$. Since it is permitted to interchange u_1 and u_2, we may suppose that $\mu_{3,1} \leq \mu_{3,2}$.

For convenience, we write $\mu_{3,1} = p$, $\mu_{3,2} = q$, and $\mu_{3,3} = r$. Letting T be the linear transformation determined by $T(u_i) = e_i$ ($i = 1, 2, 3$), we have that $T(\Lambda) = \mathbb{Z}^3$ and that $T(O)$ is the octahedron with vertices $\pm(1, 0, 0)$, $\pm(0, 1, 0)$, and $\pm(p, q, r)$. Therefore, $T(O)$ consists of the points x which satisfy

$$\left|x^1 - \frac{p}{r}x^3\right| + \left|x^2 - \frac{q}{r}x^3\right| + \left|\frac{1}{r}x^3\right| \leq 1. \tag{4.10}$$

Here, p, q, and r are integers with $0 \leq p \leq q < r$, and, by hypothesis, (4.10) is not fulfilled for any lattice point distinct from $(0, 0, 0)$, $\pm(1, 0, 0)$, $\pm(0, 1, 0)$, and $\pm(p, q, r)$. We wish to prove that $(p, q, r) = (0, 0, 1)$ or $(1, 1, 2)$. To this end, we distinguish three cases.

a. $r = 1$. Then necessarily, $p = q = 0$.

b. *$r > 1$ is an odd integer.* Then (4.10) is satisfied by $(x^1, x^2, 1)$ where

$$\begin{cases} x^1 = 0, & \text{if } p < \frac{1}{2}r; \\ x^1 = 1, & \text{if } p > \frac{1}{2}r; \\ x^2 = 0, & \text{if } q < \frac{1}{2}r; \\ x^2 = 1, & \text{if } q > \frac{1}{2}r. \end{cases}$$

c. *$r = 2s > 0$ is an even integer.* If it is not the case that $p = s$ and $q = s$, then (4.10) holds for some lattice point $x = (x^1, x^2, 1)$ where $x^1 = 0$ or 1 and $x^2 = 0$ or 1. If $p = q = s > 1$, then (4.10) holds for $x = (1, 1, 2)$. It is only when $p = q = s = 1$ that there is no new integral solution to (4.10).

Lemma 4.8 is proven. □

For convenience, we shall say that O is of type I or type II, according to whether case I or case II of the last lemma occurs. In the second case, we have

$$u_1 = v_1, \quad u_2 = v_2, \quad u_3 = \tfrac{1}{2}(-v_1 - v_2 + v_3)$$

and, thus,

$$\Lambda = \left\{\tfrac{1}{2}(\nu^1 v_1 + \nu^2 v_2 + \nu^3 v_3) : \ \nu^i \in \mathbb{Z}, \ \nu^1 \equiv \nu^2 \equiv \nu^3 \ (\text{mod } 2)\right\}. \quad (4.11)$$

Lemma 4.9 (H. Minkowski [4]). *Let O be a lattice octahedron of Λ with vertices $\pm v_i \in \partial(2C)$, where C is a centrally symmetric convex body. Then $\Lambda \in \mathcal{L}(C)$ if either*

1*. *O is of type* I *and $\theta v_i \pm v_j \pm v_k \notin \text{int}(2C)$, where $\theta = 0$, 1, 2 and $\{i, j, k\} = \{1, 2, 3\}$; or*

2*. *O is of type* II *and $\frac{1}{2}(\pm v_1 \pm v_2 \pm v_3) \notin \text{int}(2C)$.*

Proof: Suppose that 1* holds. Without loss of generality, we may suppose that $v_i = e_i$ $(i = 1, 2, 3)$. Then $\Lambda = \mathbb{Z}^3$, $e_i \in \partial(2C)$ $(i = 1, 2, 3)$, and

$$\theta e_i \pm e_j \pm e_k \notin \text{int}(2C),$$

where $\theta = 0, 1, 2$ and $\{i, j, k\} = \{1, 2, 3\}$. Assume that there exists a lattice point $u = (u^1, u^2, u^3) \neq o$ with $u \in \text{int}(2C)$. Because of symmetry, we may suppose that $0 \leq u^1 \leq u^2 \leq u^3$. Then we have $u^3 \geq 2$ and $u \neq (1, 1, 2)$. It follows from the proof of Lemma 4.8 (see cases b and c) that the octahedron O_1 with vertices $\pm e_1$, $\pm e_2$, and $\pm u$ contains some lattice point $w = (w^1, w^2, 1)$, distinct from $\pm e_1$, $\pm e_2$, and $\pm u$, with $w^i = 0$ or 1 for $i = 1, 2$. Since $2C$ is convex, $e_i \in \partial(2C)$ for $i = 1$, 2, and $u \in \text{int}(2C)$, we deduce that $w \in \text{int}(2C)$, a contradiction. Thus, we have

$$\mathbb{Z}^3 \cap \text{int}(2C) = \{o\}.$$

In other words, $\Lambda \in \mathcal{L}(C)$.

Suppose that 2* holds. Again we may suppose that $v_i = e_i$ $(i = 1, 2, 3)$. Then we have

$$e_i \in \partial(2C), \quad i = 1,\ 2,\ 3, \tag{4.12}$$

$$(\pm\tfrac{1}{2}, \pm\tfrac{1}{2}, \pm\tfrac{1}{2}) \notin \mathrm{int}(2C) \tag{4.13}$$

and, from (4.11),

$$\Lambda = \left\{ \tfrac{1}{2}(\nu^1 e_1 + \nu^2 e_2 + \nu^3 e_3) :\ \nu^i \in \mathbb{Z},\ \nu^1 \equiv \nu^2 \equiv \nu^3 \ (\mathrm{mod}\ 2) \right\}.$$

If $u = (u^1, u^2, u^3) \in \Lambda \cap \mathbb{Z}^3$, then $u \notin \mathrm{int}(2C)$. Otherwise, without loss of generality, we may assume that $0 \le u^1 \le u^2 \le u^3$, and then use (4.12) and convexity to get

$$\left(\frac{1}{2}, \frac{1}{2}, \frac{1}{2}\right) = \frac{1}{2u^3}u + \frac{u^3 - u^1}{2u^3}e_1 + \frac{u^3 - u^2}{2u^3}e_2 \in \mathrm{int}(2C),$$

which contradicts (4.13).

If $u = (u^1 + \frac{1}{2}, u^2 + \frac{1}{2}, u^3 + \frac{1}{2}) \in \Lambda \setminus \mathbb{Z}^3$, then we may again assume that $0 \le u^1 \le u^2 \le u^3$. Supposing that $u^3 \ge 1$ and denoting the octahedron with vertices $\pm e_1$, $\pm e_2$, and $\pm u$ by O_1, it is easy to see that

$$\left(\frac{1}{2}, \frac{1}{2}, \frac{1}{2}\right) = \frac{1}{2(u^3 + \frac{1}{2})}u + \beta e_1 + \gamma e_2$$

holds with $\beta \ge 0$, $\gamma \ge 0$, and $\frac{1}{2(u^3+\frac{1}{2})} + \beta + \gamma \le 1$, and then $(\frac{1}{2}, \frac{1}{2}, \frac{1}{2}) \in \mathrm{int}(O_1)$. Hence, $u \notin \mathrm{int}(2C)$.

In conclusion, we see that Λ is also a packing lattice of C in case 2*, and so Lemma 4.9 is proven. □

Remark 4.1. *If the convex body C of Lemma 4.9 is regular, then it is easy to see from the proof that $\pm v_1$, $\pm v_2$, $\pm v_3$, and the points mentioned in the corresponding cases are the only candidates for points v in Λ for which*

$$C \cap (v + C) \ne \emptyset.$$

Lemma 4.10 (H. Minkowski [4]). *For every centrally symmetric convex body C in R^3, there is a $\Lambda \in \mathcal{L}^\delta(C)$ with a basis $\{u_1, u_2, u_3\}$ such that either*

$$\{u_1, u_2, u_3, u_1 - u_2, u_1 - u_3, u_2 - u_3\} \subset \partial(2C)$$

or

$$\{u_1, u_2, u_3, u_1 + u_2, u_1 + u_3, u_2 + u_3\} \subset \partial(2C).$$

Proof: By Lemma 4.4 and Mahler's selection theorem, we may assume, without loss of generality, that C is regular. We begin by choosing an arbitrary $\Lambda \in \mathcal{L}^\delta(C)$ and consider the following two cases.

Case 1. *Each triple of independent points of $\Lambda \cap \partial(2C)$ determines a lattice octahedron of type* I.

Let $\{v_1, v_2, v_3\}$ be such a triple and take

$$X = \{\nu_1 v_1 + \nu_2 v_2 + \nu_3 v_3 \neq o : \ \nu_i = -1, 0, 1\}.$$

From the assumption of case 1, we notice that

$$\pm 2v_i \pm v_j \pm v_k \notin \partial(2C), \quad \{i, j, k\} = \{1, 2, 3\}.$$

Therefore, by Remark 4.1, we get $\Lambda \cap \partial(2C) \subset X$. Now we prove

Assertion 4.1. *If v_1, v_2, and v_3 are three points of $\Lambda \cap \partial(2C)$, then there exists a point $w \in X \cap \partial(2C)$ which does not belong to*

$$Y = \{\pm v_3, \ \pm(v_1 + \nu_1 v_3), \ \pm(v_2 + \nu_2 v_3) : \ \nu_1, \ \nu_2 \in \mathbb{Z}\}.$$

Since $\{v_2, v_1 + v_3, v_1 - v_3\}$ is not a basis of Λ, by the assumption of case 1, $v_1 + v_3$ and $v_1 - v_3$ cannot simultaneously belong to $\Lambda \cap \partial(2C)$. Thus, if $X \cap \partial(2C) \subset Y$, we get

$$\begin{aligned} h(C, \Lambda) &= \text{card}\{\Lambda \cap \partial(2C)\} \le \text{card}\{X \cap \partial(2C)\} \\ &\le \text{card}\{Y \cap \partial(2C)\} \le 10, \end{aligned}$$

which contradicts Lemma 4.7. Thus, Assertion 4.1 is true.

Keeping Assertion 4.1 and the assumption of Case 1 in mind, by considering various cases we see that there exists a triple of points u_1, u_2, and u_3 such that

$$\{u_1, u_2, u_3, u_1 - u_2, u_1 - u_3, u_2 - u_3\} \subset \partial(2C).$$

Case 2. *There are three independent points in $\Lambda \cap \partial(2C)$ determining a lattice octahedron of the type* II.

Let v_1, v_2, and v_3 be such a triple and take

$$X' = \left\{\pm \tfrac{1}{2} v_1 \pm \tfrac{1}{2} v_2 \pm \tfrac{1}{2} v_3\right\}.$$

Divide X' into four groups, $\{-\frac{1}{2}v_1 - \frac{1}{2}v_2 - \frac{1}{2}v_3, \frac{1}{2}v_1 + \frac{1}{2}v_2 + \frac{1}{2}v_3\}$, $\{-\frac{1}{2}v_1 + \frac{1}{2}v_2 + \frac{1}{2}v_3, \frac{1}{2}v_1 - \frac{1}{2}v_2 - \frac{1}{2}v_3\}$, $\{\frac{1}{2}v_1 - \frac{1}{2}v_2 + \frac{1}{2}v_3, -\frac{1}{2}v_1 + \frac{1}{2}v_2 - \frac{1}{2}v_3\}$, and $\{\frac{1}{2}v_1 + \frac{1}{2}v_2 - \frac{1}{2}v_3, -\frac{1}{2}v_1 - \frac{1}{2}v_2 + \frac{1}{2}v_3\}$. For convenience, we abbreviate them as $\{u_1, -u_1\}$, $\{u_2, -u_2\}$, $\{u_3, -u_3\}$, and $\{u_4, -u_4\}$, respectively. By Lemma 4.7 and Remark 4.1, we find that at least three of these four groups belong to $\partial(2C)$. Without loss of generality, we may suppose that

$$u_i \in \partial(2C), \quad i = 1, \ 2, \ 3.$$

Then it can be easily shown that

$$u_1 + u_2, u_1 + u_3, u_2 + u_3 \in \partial(2C).$$

Lemma 4.10 is thus proven. □

§5. A Phenomenon Concerning Kissing Numbers and Packing Densities

Phenomenon β: *Let S be a tetrahedron. Then there exist two lattices $\Lambda_1 \in \mathcal{L}^\delta(S)$ and $\Lambda_2 \in \mathcal{L}^h(S)$ such that*

$$\delta(S,\Lambda_1) = \delta^*(S) = \tfrac{18}{49}, \quad h(S,\Lambda_1) = 14$$

and

$$\delta(S,\Lambda_2) = \tfrac{1}{3}, \quad h(S,\Lambda_2) = h^*(S) = 18.$$

For convenience, we separate the verification of this phenomenon into two theorems.

Theorem 4.1 (D.J. Hoylman [1]).[2] *Let S be a tetrahedron. Then $\delta^*(S) = \frac{18}{49}$. Moreover, there exists a $\Lambda \in \mathcal{L}^\delta(S)$ such that $h(S,\Lambda) = 14$.*

Sketch of the Proof: We deal with $\frac{1}{2}D(S)$ instead of S. We consider all the lattices meeting the conditions of Lemma 4.10 such that none of the lattice points listed in Lemma 4.9 belongs to $\text{int}(D(S))$ and find the one with the smallest determinant.

Let S be the particular tetrahedron with vertices $(-1,1,1)$, $(1,-1,1)$, $(1,1,-1)$, and $(-1,-1,-1)$. It can be shown that

$$D(S) = \{(x^1,x^2,x^3) : \ |x^i| \le 2, \ |x^1| + |x^2| + |x^3| \le 4\}.$$

Geometrically speaking, $D(S)$ is that polyhedon with six square and eight triangular facets known as the *cuboctahedron.* The proof may now be divided into cases according to the distribution of the basis vectors u_1, u_2, and u_3 among the facets of $D(S)$. It can be easily shown that no triangular facet of $D(S)$ can contain two points of a packing lattice of $\frac{1}{2}D(S)$ in its interior. Using this fact and the symmetries of $D(S)$, we find 19 essentially different ways to assign u_1, u_2, and u_3 to the facets of $D(S)$. Combining each of these with the 2 cases of Lemma 4.10, we obtain 38 cases. Computing the minimal value of the determinant of the lattices in question for each of these cases, it can be shown that in none of these cases can we obtain a lattice meeting the above conditions whose determinant is less than $\frac{196}{27}$. This value is the determinant of the lattice $\Lambda \in \mathcal{L}(\frac{1}{2}D(S))$ with basis

$$u_1 = \left(2, -\tfrac{1}{3}, -\tfrac{1}{3}\right), \quad u_2 = \left(-\tfrac{1}{3}, 2, -\tfrac{1}{3}\right), \quad u_3 = \left(-\tfrac{1}{3}, -\tfrac{1}{3}, 2\right).$$

---

[2] Hoylman's article also included a description of $\mathcal{L}^\delta(S)$. However, since Lemma 4.10 was improperly used, there was a gap in his proof of this description of $\mathcal{L}^\delta(S)$. For more information, we refer to the remarks at the end of this chapter.

Since $v(S) = \frac{8}{3}$, we get

$$\delta^*(S) = \tfrac{18}{49}.$$

The verification that $h(S, \Lambda) = 14$ is routine and thus omitted here. □

Theorem 4.2 (C. Zong [5]). *Let S be a tetrahedron. Then $h^*(S) = 18$. Moreover, there exists a $\Lambda \in \mathcal{L}^h(S)$ such that $\delta(S, \Lambda) = \frac{1}{3}$.*

Before we can begin the complicated proof, some notation and definitions are necessary.

Let $abcd$ denote the quadrilateral with vertices a, b, c, and d, i.e., the convex hull of $\{a, b, c, d\}$; let abc denote the triangle with vertices a, b, and c, i.e., the convex hull of $\{a, b, c\}$; let ab denote the closed linear segment from a to b; let $[ab)$ denote the half-closed linear segment from a to b that contains a; and let (ab) denote the open linear segment from a to b.

Denoting the tetrahedron with vertices $p_1 = o = (0, 0, 0)$, $p_2 = (1, 0, 0)$, $p_3 = (0, 1, 0)$, and $p_4 = (0, 0, -1)$ by S, let $S + T$ be a translative packing of S in R^3 (thus, T is a set of points $t_i \in R^3$ with $t_0 = o$). We write S_i for the translated tetrahedra $S + t_i$ and $p_{j,i}$ for the translated vertices $p_j + t_i$ (so, in particular, $S_0 = S$ and $p_{j,0} = p_j$). Denote the distance from p to the straight line determined by x and y by $d(p, \overline{xy})$. Let $M(S)$ be the number of translates S_i from $S + T$ with $i \neq 0$ which are in contact with S and let $I(p_1p_2p_3)$ be the number of nonzero indices i such that $S_i \cap p_1p_2p_3 = \{p\}$ and $p \in \operatorname{int}(p_1p_2p_3)$. (When we speak of the interior of a facet or an edge, we always mean the relative interior in the corresponding two- or one-dimensional space, respectively.) With these preliminaries behind us, we are now ready to define a valuation function. Let

$$Z(S_i, p_1p_2p_3) =$$

$$\begin{cases} \frac{1}{3} & \text{if } S_i \cap p_1p_2p_3 = \{p\} \text{ and } p \text{ is an interior point of a facet of } S_i \\ \frac{1}{2} & \text{if } S_i \cap p_1p_2p_3 = \{p\} \text{ and } p \text{ is the midpoint of } p_{4,i}p_{j,i} \text{ for some } j \neq 4 \\ 1 & \text{if } S_i \cap p_1p_2p_3 = \{p\},\ p \in p_{4,i}p_{j,i} \text{ for some } j, \text{ and } \|p_{j,i}p\| > \|p_{4,i}p\| \\ 0 & \text{in all other cases (especially, when } i = 0), \end{cases}$$

and then set

$$G(p_1p_2p_3) = \sum_{S_i \cap S \neq \emptyset} Z(S_i, p_1p_2p_3).$$

For convenience, we abbreviate $\sum Z(S_i, p_1p_2p_3)$ by $\sum$.

Remark 4.2. *We define $Z(S_i, p_1p_2p_4)$, $Z(S_i, p_1p_3p_4)$, $Z(S_i, p_2p_3p_4)$, $G(p_1p_2p_4)$, $G(p_1p_3p_4)$, and $G(p_2p_3p_4)$ in a similar manner.*

Remark 4.3. *From the definition of $Z(S_i, p_1p_2p_3)$, we see the following:*

a. *If $Z(S_i, p_1p_2p_3) = \frac{1}{2}$ and $S_i \cap p_1p_2p_3 = \{p\}$, then $p + (0, 0, \frac{1}{2}) + \frac{1}{2}S \subseteq S_i$.*

b. *If $Z(S_i, p_1p_2p_3) = 1$ and $S_i \cap p_1p_2p_3 = \{p\}$, then a positive number η exists such that $p + (0, 0, \frac{1}{2} + \eta) + (\frac{1}{2} + \eta)S \subseteq S_i$.*

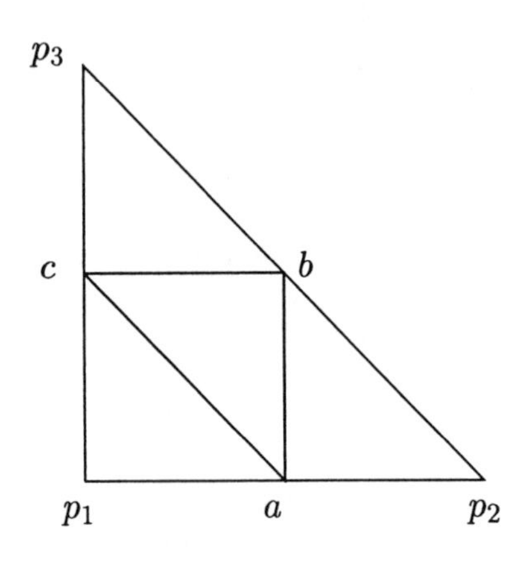

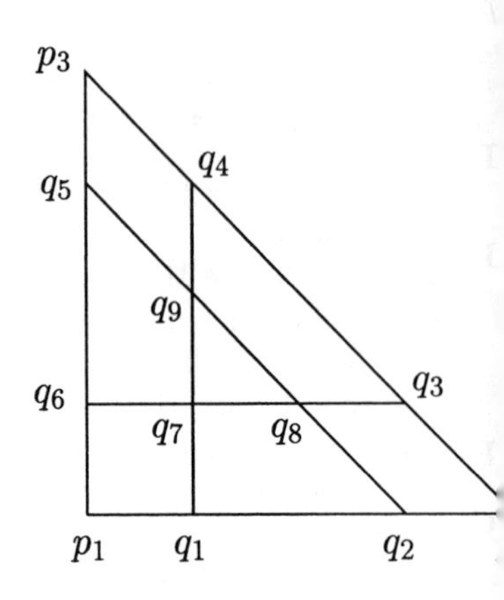

Figure 7

Let a, b, and c be the midpoints of p_1p_2, p_2p_3, and p_3p_1, respectively, let $q_1, q_2 \in p_1p_2$, $q_3, q_4 \in p_2p_3$, and $q_5, q_6 \in p_3p_1$ be the points such that

$$\frac{p_1q_1}{p_1p_2} = \frac{q_2p_2}{p_1p_2} = \frac{p_2q_3}{p_2p_3} = \frac{q_4p_3}{p_2p_3} = \frac{p_3q_5}{p_3p_1} = \frac{q_6p_1}{p_3p_1} = \frac{1}{4},$$

and let $q_7 = q_1q_4 \cap q_3q_6$, $q_8 = q_2q_5 \cap q_3q_6$, and $q_9 = q_2q_5 \cap q_1q_4$. Finally, define

$$ap_2b^- = ap_2b \setminus \{b\}, \quad bp_3c^- = bp_3c \setminus \{c\}, \quad cp_1a^- = cp_1a \setminus \{a\},$$

$$\begin{aligned} H_1(p_1p_2p_3) &= \{p_1q_1q_7q_6 \setminus (q_1q_7 \cup q_6q_7)\} \cup \{q_2p_2q_3q_8 \setminus (q_3q_8 \cup q_2q_8)\} \\ &\quad \cup \{p_3q_5q_9q_4 \setminus (q_5q_9 \cup q_4q_9)\}, \\ H_2(p_1p_2p_3) &= \{q_1q_2q_8q_7 \setminus q_7q_8\} \cup \{q_3q_4q_9q_8 \setminus q_8q_9\} \cup \{q_5q_6q_7q_9 \setminus q_7q_9\}, \\ H_3(p_1p_2p_3) &= q_7q_8q_9. \end{aligned}$$

Remark 4.4. *The $H_k(p_1p_2p_3)$ have the following properties.*

a. $p_1p_2p_3 = H_1(p_1p_2p_3) \cup H_2(p_1p_2p_3) \cup H_3(p_1p_2p_3)$.

b. $H_i(p_1p_2p_3) \cap H_j(p_1p_2p_3) = \emptyset$ *for* $i \neq j$.

c. *For every point $x \in H_k(p_1p_2p_3)$, there are k edges p_ip_j of $p_1p_2p_3$ such that $d(x, \overline{p_ip_j}) \geq \frac{1}{4}d(p_l, \overline{p_ip_j})$, where $\{l, i, j\} = \{1, 2, 3\}$.*

Before turning to some complicated estimations, we state two intersection properties of translated tetrahedra which will be used frequently.

Assertion 4.2. *If* $\mathrm{card}\{S \cap (S+p)\} \geq 2$, *then* $\mathrm{int}(S) \cap (\mathrm{int}(S)+p) \neq \emptyset$.

Assertion 4.3. *If* $x, y \in z + \eta\{p_1p_2p_3\}$ *for some* z *and* η, *then for any* $\rho > \eta$,

$$(x + \mathrm{int}(\rho S)) \cap (y + \mathrm{int}(\rho S)) \neq \emptyset.$$

Remark 4.5. *From Assertion* 4.3, *it follows that* $I(p_1p_2p_3) \leq 1$.

Lemma 4.11.

$$M(S) = G(p_1p_2p_3) + G(p_1p_2p_4) + G(p_1p_3p_4) + G(p_2p_3p_4).$$

Proof: For $i \neq 0$, exactly one of the following cases must hold:

Case 1. $S_i \cap S = \emptyset$.

Clearly, we then have that

$$\begin{cases} Z(S_i, p_1p_2p_3) & = 0, \\ Z(S_i, p_1p_2p_4) & = 0, \\ Z(S_i, p_1p_3p_4) & = 0, \\ Z(S_i, p_2p_3p_4) & = 0. \end{cases} \tag{4.14}$$

Case 2. $S_i \cap S = \{p\}$, *where* p *is an interior point of a facet of* S_i.

Without loss of generality, we take $p_{2,i}p_{3,i}p_{4,i}$ to be the facet under consideration. Then, since $S+T$ is a translative packing, we see that $p = p_1$ and that

$$\begin{cases} Z(S_i, p_1p_2p_3) & = \frac{1}{3}, \\ Z(S_i, p_1p_2p_4) & = \frac{1}{3}, \\ Z(S_i, p_1p_3p_4) & = \frac{1}{3}, \\ Z(S_i, p_2p_3p_4) & = 0. \end{cases} \tag{4.15}$$

Case 3. $S_i \cap S = \{p\}$, *where* p *is an interior point of an edge of* S_i.

Without loss of generality, we take $p_{2,i}p_{3,i}$ to be the edge under consideration. Keeping in mind the intersection properties of our translated tetrahedra and the fact that $S+T$ is a packing, it is easy to see that $p \in p_1p_4$. Thus, if $\|pp_{2,i}\| = \|pp_{3,i}\|$, then

$$\begin{cases} Z(S_i, p_1p_2p_3) & = 0, \\ Z(S_i, p_1p_2p_4) & = \frac{1}{2}, \\ Z(S_i, p_1p_3p_4) & = \frac{1}{2}, \\ Z(S_i, p_2p_3p_4) & = 0; \end{cases} \tag{4.16}$$

whereas if $\|pp_{2,i}\| > \|pp_{3,i}\|$, then

$$\begin{cases} Z(S_i, p_1p_2p_3) & = 0, \\ Z(S_i, p_1p_2p_4) & = 1, \\ Z(S_i, p_1p_3p_4) & = 0, \\ Z(S_i, p_2p_3p_4) & = 0. \end{cases} \tag{4.17}$$

Case 4. *$S_i \cap S = \{p\}$, where p is a vertex of S_i.*

Without loss of generality, we take $p = p_{4,i}$ to be the vertex under consideration. Then, we see that p is a point of $p_1p_2p_3$ and that

$$\begin{cases} Z(S_i, p_1p_2p_3) & = 1, \\ Z(S_i, p_1p_2p_4) & = 0, \\ Z(S_i, p_1p_3p_4) & = 0, \\ Z(S_i, p_2p_3p_4) & = 0. \end{cases} \tag{4.18}$$

From (4.14)–(4.18), we conclude that

$$Z(S_i, p_1p_2p_3) + Z(S_i, p_1p_2p_4) + Z(S_i, p_1p_3p_4) + Z(S_i, p_2p_3p_4)$$

$$= \begin{cases} 0 & \text{if } S \cap S_i = \emptyset \\ 0 & \text{if } i = 0 \\ 1 & \text{if } S \cap S_i \neq \emptyset \text{ and } i \neq 0. \end{cases}$$

Hence,

$$\begin{aligned} M(S) &= \sum_{S_i \cap S \neq \emptyset,\ i \neq 0} 1 \\ &= \sum_{S_i \cap S \neq \emptyset} Z(S_i, p_1p_2p_3) + \sum_{S_i \cap S \neq \emptyset} Z(S_i, p_1p_2p_4) \\ &\quad + \sum_{S_i \cap S \neq \emptyset} Z(S_i, p_1p_3p_4) + \sum_{S_i \cap S \neq \emptyset} Z(S_i, p_2p_3p_4) \\ &= G(p_1p_2p_3) + G(p_1p_2p_4) + G(p_1p_3p_4) + G(p_2p_3p_4), \end{aligned}$$

and so Lemma 4.11 is proven. □

Lemma 4.12. *If $I(p_1p_2p_3) = 0$, then*

$$G(p_1p_2p_3) \leq 4\tfrac{2}{3}.$$

Proof: From our definitions, we see that in order to estimate $G(p_1p_2p_3)$, one must carefully estimate

$$\sum_{S_i \cap ap_2b^- \neq \emptyset}, \quad \sum_{S_i \cap bp_3c^- \neq \emptyset}, \quad \text{and} \quad \sum_{S_i \cap cp_1a^- \neq \emptyset}.$$

From Remark 4.3, Assertion 4.3, and the asssumption that $S+T$ is a packing, we can easily deduce the following:

Assertion 4.4. *At most one translate S_i exists such that $S_i \cap [ap_2) \neq \emptyset$ and $Z(S_i, p_1p_2p_3) \geq \frac{1}{2}$.*

Assertion 4.5. *At most one translate S_i exists such that $S_i \cap ap_2b^- \neq \emptyset$ and $Z(S_i, p_1p_2p_3) = 1$.*

Assertion 4.6. *If a translate S_j exists such that $S_j \cap (ap_2) \neq \emptyset$ and $Z(S_j, p_1p_2p_3) \neq 0$, then $\sum_{S_i \cap (p_2b) \neq \emptyset} = 0$.*

From Assertions 4.4–4.6 and the assumption that $I(p_1p_2p_3) = 0$, we conclude that

$$\sum_{S_i \cap ap_2b^- \neq \emptyset} = l + \frac{m}{2} + \frac{n}{3}, \tag{4.19}$$

where $l, m, n \in \{0, 1, 2\}$, $l \leq 1$, $l + m \leq 2$, and $n \leq 1$. Obviously, (4.19) holds for $\sum_{S_i \cap bp_3c^- \neq \emptyset}$ and $\sum_{S_i \cap cp_1a^- \neq \emptyset}$ also. Setting

$$M = \max \left\{ \sum_{S_i \cap ap_2b^- \neq \emptyset}, \sum_{S_i \cap bp_3c^- \neq \emptyset}, \sum_{S_i \cap cp_1a^- \neq \emptyset} \right\},$$

we may assume, without loss of generality, that $M = \sum_{S_i \cap ap_2b^- \neq \emptyset}$.

From (4.19), we see that

$$M \leq 1 + \tfrac{1}{2} + \tfrac{1}{3}.$$

Our proof is completed by dealing with the following cases:

Case 1. $M = 1 + \frac{1}{2} + \frac{1}{3}$.

Recalling Assertions 4.4–4.6, we see that in this case there must exist two translates S_i and S_j such that

$$\begin{cases} S_i \cap S & = \{\text{a}\}, \\ Z(S_i, p_1p_2p_3) & \geq \frac{1}{2} \end{cases}$$

and

$$\begin{cases} S_j \cap S & = \{p\}, \ \text{where } p \in (p_2b), \\ Z(S_j, p_1p_2p_3) & \geq \frac{1}{2}. \end{cases}$$

Then, applying Assertions 4.4–4.6 to bp_3c^- and cp_1a^-, we deduce that

$$\begin{cases} \sum_{S_i \cap bp_3c^- \neq \emptyset} & \leq 1 + \frac{1}{3}, \\ \sum_{S_i \cap cp_1a^- \neq \emptyset} & \leq 1 + \frac{1}{2}. \end{cases}$$

Keeping in mind the assumption that $I(p_1p_2p_3) = 0$, it follows that

$$G(p_1p_2p_3) = \sum_{S_i \cap ap_2b^- \neq \emptyset} + \sum_{S_i \cap bp_3c^- \neq \emptyset} + \sum_{S_i \cap cp_1a^- \neq \emptyset} \leq 4\tfrac{2}{3}.$$

Case 2. $M < 1 + \frac{1}{2} + \frac{1}{3}$.

In this case, from (4.19) we see that $M \leq 1 + \frac{1}{2}$. Then, from the definition of M and the assumption that $I(p_1p_2p_3) = 0$, it follows that

$$G(p_1p_2p_3) = \sum_{S_i \cap ap_2b^- \neq \emptyset} + \sum_{S_i \cap bp_3c^- \neq \emptyset} + \sum_{S_i \cap cp_1a^- \neq \emptyset} \leq 4\tfrac{1}{2}.$$

With this, Lemma 4.12 is proven. □

Lemma 4.13. *If* $I(p_1p_2p_3) = 1$, *then*

$$G(p_1p_2p_3) \leq 5,$$

with equality being attained only when a translate S_i *exists such that* $S_i \cap p_1p_2p_3 = \{p\}$, *where* $p \in \text{int}(abc)$.

Proof: We show our lemma by proving the following assertions:

Assertion 4.7. *If a translate* S_i *exists such that*

$$\begin{cases} S_i \cap p_1p_2p_3 & = \{p\}, \text{ where } p \in \text{int}(p_1p_2p_3) \setminus \text{int}(abc), \\ Z(S_i, p_1p_2p_3) & = 1, \end{cases}$$

then

$$G(p_1p_2p_3) \leq 4\tfrac{2}{3}.$$

Without loss of generality, we may assume that

$$p \in ap_2b \setminus \{ap_2 \cup p_2b\}.$$

Then, from Remark 4.3, Assertion 4.3, and the assumption that $S + T$ is a packing, we see that at most one other translate S_j exists such that $S_j \cap ap_2b \neq \emptyset$ and $Z(S_j, p_1p_2p_3) > 0$; moreover, for that S_j,

$$Z(S_j, p_1p_2p_3) \leq \tfrac{1}{2}.$$

In consequence,

$$\sum_{S_i \cap ap_2b^- \neq \emptyset} \leq 1 + \tfrac{1}{2}. \tag{4.20}$$

At the same time, applying Assertions 4.4–4.6 to cp_1a^- and bp_3c^-, we see that

$$\begin{cases} \sum_{S_i \cap cp_1a^- \neq \emptyset} & \leq 1 + \frac{1}{2} + \frac{1}{3}, \\ \sum_{S_i \cap bp_3c^- \neq \emptyset} & \leq 1 + \frac{1}{3}\ . \end{cases} \tag{4.21}$$

From (4.20) and (4.21), we conclude that

$$G(p_1p_2p_3) = \sum_{S_i \cap ap_2b^- \neq \emptyset} + \sum_{S_i \cap bp_3c^- \neq \emptyset} + \sum_{S_i \cap cp_1a^- \neq \emptyset} \leq 4\tfrac{2}{3}.$$

Assertion 4.8. *If a translate S_i exists such that*

$$\begin{cases} S_i \cap \operatorname{int}(abc) & \neq \emptyset, \\ Z(S_i, p_1p_2p_3) & = 1, \end{cases}$$

then

$$G(p_1p_2p_3) \leq 5.$$

In this case, there is no translate S_j such that

$$\begin{cases} S_j \cap p_1p_2p_3 & = \{p\}, \text{ where } p \in \{a, b, c\}, \\ Z(S_j, p_1p_2p_3) & > 0 \end{cases}$$

(otherwise, by applying Remark 4.3 and Assertion 4.2, with some detailed argument it can be shown that $S + T$ is not a packing). Then, applying Assertions 4.4–4.6 to ap_2b^-, bp_3c^-, and cp_1a^-, we see that

$$\begin{cases} \sum_{S_i \cap ap_2b^- \neq \emptyset} & \leq 1 + \frac{1}{3}, \\ \sum_{S_i \cap bp_3c^- \neq \emptyset} & \leq 1 + \frac{1}{3}, \\ \sum_{S_i \cap cp_1a^- \neq \emptyset} & \leq 1 + \frac{1}{3}\ . \end{cases} \tag{4.22}$$

From (4.22) and the assumption that there is a translate S_i such that $S_i \cap \operatorname{int}(abc) \neq \emptyset$, we conclude that

$$G(p_1p_2p_3) = 1 + \sum_{S_i \cap ap_2b^- \neq \emptyset} + \sum_{S_i \cap bp_3c^- \neq \emptyset} + \sum_{S_i \cap cp_1a^- \neq \emptyset} \leq 5.$$

The lemma now follows immediately from Assertions 4.7 and 4.8. □

Lemma 4.14. *Suppose that p_2 and p_3 are interior points of $p_{1,2}p_{3,2}p_{4,2}$ and $p_{1,3}p_{2,3}p_{4,3}$, respectively. If*

$$\begin{cases} d(p_2, \overline{p_{1,2}p_{3,2}}) & \geq \frac{1}{4} d(p_{4,2}, \overline{p_{1,2}p_{3,2}}), \\ d(p_3, \overline{p_{1,3}p_{2,3}}) & \geq \frac{1}{4} d(p_{4,3}, \overline{p_{1,3}p_{2,3}}), \end{cases} \tag{4.23}$$

then

$$G(p_1p_2p_3) < 5.$$

Proof: If, on the contrary, $G(p_1p_2p_3) = 5$, then by Lemma 4.13 and its proof, a translate S_j exists such that

$$S_j \cap S = \{p_{4,j}\}, \text{ where } p_{4,j} = (x^1, x^2, 0) \in \operatorname{int}(abc)$$

and

$$\begin{cases} \sum_{S_j \cap \{(ap_2) \cup [p_2 b)\} \neq \emptyset} & = 1 + \frac{1}{3}, \\ \sum_{S_j \cap \{(bp_3) \cup [p_3 c)\} \neq \emptyset} & = 1 + \frac{1}{3}. \end{cases} \tag{4.24}$$

From (4.24) and Assertion 4.6, we obtain two points y and z and two translates S_k and S_l such that

$$\begin{cases} y = (y^1, y^2, 0) \in S_k \cap \{(ap_2) \cup (p_2 b)\}, \\ Z(S_k, p_1 p_2 p_3) = 1 \end{cases}$$

and

$$\begin{cases} z = (z^1, z^2, 0) \in S_l \cap \{(bp_3) \cup (p_3 c)\}, \\ Z(S_l, p_1 p_2 p_3) = 1. \end{cases}$$

Thus, from Remark 4.3 we see that for some positive number η,

$$\begin{cases} y + (0, 0, \frac{1}{2} + \eta) + (\frac{1}{2} + \eta)S & \subseteq S_k, \\ z + (0, 0, \frac{1}{2} + \eta) + (\frac{1}{2} + \eta)S & \subseteq S_l. \end{cases} \tag{4.25}$$

Then, since (4.23) implies that

$$\begin{cases} p_2 + \left(0, 0, \frac{1}{4}\right) + \frac{1}{4}S & \subseteq S_2, \\ p_3 + \left(0, 0, \frac{1}{4}\right) + \frac{1}{4}S & \subseteq S_3, \end{cases}$$

by keeping in mind Assertion 4.3 and the assumption that $S + T$ is a packing, we see that

$$\begin{cases} y = (y^1, y^2, 0) \in (ap_2) \cup (p_2 b), \\ \frac{1}{2} < y^1 \leq \frac{3}{4} \end{cases} \tag{4.26}$$

and

$$\begin{cases} z = (z^1, z^2, 0) \in (bp_3) \cup (p_3 c), \\ \frac{1}{2} < z^2 \leq \frac{3}{4}. \end{cases} \tag{4.27}$$

We now finish the proof of this lemma by disposing of the following cases.

Case 1. $y \in ap_2$, $z \in p_3 c$.

Since $p_{4,j} = (x^1, x^2, 0) \in \text{int}(abc)$, at least one of the two coordinates x^1 and x^2 is greater than $\frac{1}{4}$. Without loss of generality, let $x^1 > \frac{1}{4}$. Keeping (4.25) and (4.26) in mind, we then see that both $(y^1, x^2, \frac{1}{2})$ and $(y^1, x^2, \frac{1}{2} + \eta)$ belong to $S_j \cap S_k$. Hence, from Assertion 4.2 we have that

$$\text{int}(S_j) \cap \text{int}(S_k) \neq \emptyset,$$

which is contrary to the assumption that $S + T$ is a packing.

Case 2. $y \in ap_2$, $z \in bp_3$.

When $x^1 > \frac{1}{4}$, the argument of case 1 can be repeated. When $x^2 > \frac{1}{4}$, it can be verified that both $z + (0, 0, \frac{1}{2})$ and $z + (0, 0, \frac{1}{2} + \eta)$ belong to $S_j \cap S_l$. Hence, from Assertion 4.2 we have that

$$\operatorname{int}(S_j) \cap \operatorname{int}(S_l) \neq \emptyset,$$

which is contrary to the assumption that $S + T$ is a packing.

Case 3. $y \in p_2b$, $z \in p_3c$.

This case is handeled just as case 2 was.

Case 4. $y \in p_2b$, $z \in bp_3$.

We proceed as in case 1. Without loss of generality, let $x^1 > \frac{1}{4}$. Keeping (4.25) and (4.26) in mind, we then see that both $(y^1, x^2, \frac{1}{2})$ and $(y^1, x^2, \frac{1}{2} + \eta)$ belong to $S_j \cap S_k$. Hence, from Assertion 4.2 we have that

$$\operatorname{int}(S_j) \cap \operatorname{int}(S_k) \neq \emptyset,$$

which is contrary to the assumption that $S + T$ is a packing.

Since cases 1–4 all lead to a contradiction, Lemma 4.14 is proven. □

Proof of Theorem 4.2: Lemmas 4.11–4.13 show that $M(S) \leq 20$. We now prove that $M(S) \neq 20$.

If, on the contrary, $M(S) = 20$ for some T, then by Lemmas 4.12 and 4.13, we see that

$$G(p_1p_2p_3) = G(p_1p_2p_4) = G(p_1p_3p_4) = G(p_2p_3p_4) = 5.$$

Hence, indices $i(1)$, $i(2)$, $i(3)$, $i(4)$, $j(1)$, $j(2)$, $j(3)$, and $j(4)$ exist such that

$$\begin{cases} S_{i(k)} \cap S &= \{p_k\} \text{ with } p_k \in \operatorname{int}(p_{l,i(k)}p_{m,i(k)}p_{n,i(k)}), \\ S_{j(k)} \cap S &= \{p_{k,j(k)}\} \text{ with } p_{k,j(k)} \in \operatorname{int}(p_lp_mp_n), \end{cases}$$

where k, l, m, and n are such that $\{k, l, m, n\} = \{1, 2, 3, 4\}$.

Define

$$\psi(p_k, p_kp_mp_n) = \psi(p_k, p_mp_kp_n) = \psi(p_k, p_mp_np_k)$$

$$= \begin{cases} 0 & \text{if } d(p_k, \overline{p_{m,i(k)}p_{n,i(k)}}) < \frac{1}{4}d(p_{l,i(k)}, \overline{p_{m,i(k)}p_{n,i(k)}}) \\ 1 & \text{if } d(p_k, \overline{p_{m,i(k)}p_{n,i(k)}}) \geq \frac{1}{4}d(p_{l,i(k)}, \overline{p_{m,i(k)}p_{n,i(k)}}) \end{cases}$$

and

$$\Psi(p_kp_mp_n) = \psi(p_k, p_kp_mp_n) + \psi(p_m, p_kp_mp_n) + \psi(p_n, p_kp_mp_n).$$

Recalling to mind the sets $H_i(p_1p_2p_3)$ defined after Remark 4.3, let us suppose that

$$p_k \in H_{\phi(k)}(p_{l,i(k)}p_{m,i(k)}p_{n,i(k)}).$$

Then, from Remark 4.4, we see that

$$\Psi(p_1p_2p_3) + \Psi(p_1p_2p_4) + \Psi(p_1p_3p_4) + \Psi(p_2p_3p_4) = \sum_{k=1}^{4} \phi(k).$$

So if

$$\sum_{k=1}^{4} \phi(k) > 4,$$

then there exists a triangle, say $p_1p_2p_3$, such that

$$\Psi(p_1p_2p_3) \geq 2,$$

which implies an inequality like (4.23). This contradicts $G(p_1p_2p_3) = 5$; thus, we have

$$\sum_{k=1}^{4} \phi(k) = 4.$$

But then $\phi(k) = 1$ for $1 \leq k \leq 4$, and therefore

$$p_k \in H_1(p_{l,i(k)}p_{m,i(k)}p_{n,i(k)}), \quad k = 1,\ 2,\ 3,\ 4. \tag{4.28}$$

It is easy to see that $H_1(p_{1,i(2)}p_{3,i(2)}p_{4,i(2)})$ consists of three nonoverlapping parallelograms each of which contains a vertex of $p_{1,i(2)}p_{3,i(2)}p_{4,i(2)}$. Consider the case $k = 2$ of (4.28). If p_2 is a point of that parallelogrammic component of $H_1(p_{1,i(2)}p_{3,i(2)}p_{4,i(2)})$ which contains $p_{4,i(2)}$, then

$$d(p_2, \overline{p_{1,i(2)}p_{3,i(2)}}) > \tfrac{1}{2} d(p_{4,i(2)}, \overline{p_{1,i(2)}p_{3,i(2)}}).$$

This implies that for some positive η,

$$\left(1, 0, \tfrac{1}{2} + \eta\right) + \left(\tfrac{1}{2} + \eta\right) S \subseteq S_{i(2)}.$$

Then, by applying Assertion 4.3 and engaging in some routine argumentation, we can show that

$$\sum_{S_i \cap ap_2b^- \neq \emptyset} \leq \tfrac{1}{3},$$

and hence

$$G(p_1p_2p_3) < 5.$$

With this contradiction we have shown that

$$h(S) \leq 19.$$

Moreover, since $h^*(S)$ cannot be an odd number and $h^*(S) \leq h(S)$, we obtain

$$h^*(S) \leq 18. \tag{4.29}$$

To show that equality holds in (4.29), let S be the tetrahedron with vertices $(0,0,0)$, $(1,0,0)$, $(0,1,0)$, and $(0,0,1)$, and let Λ be the lattice having $\frac{1}{2}(-1,1,1)$, $\frac{1}{2}(1,-1,1)$, and $\frac{1}{2}(1,1,-1)$ as a basis. It is routine to verify that $S+\Lambda$ is a lattice packing and that every tetrahedron in this packing is in contact with 18 others. In consequence, $h^*(S)=18$. Moreover, $\delta(S,\Lambda)=\frac{1}{3}$. Thus, Theorem 4.2 is proven. □

§6. Remarks and Open Problems

In 1611, J. Kepler [1] formulated the following famous problem.

The Sphere Packing Problem: *Determine the density of the densest sphere packing in R^3.*

Toward the end of the seventeenth century, D. Gregory and I. Newton in conversation raised another well-known problem conerning sphere packing.

The Thirteen Sphere Problem: *Can a rigid material sphere be brought into contact with 13 other such spheres of the same size?*

During the last three centuries, these problems and their generalizations have been studied by many prominent mathematicians, including Lagrange, Gauss, Dirichlet, Minkowski, and others. Recently, W.Y. Hsiang [1] has announced a solution to the sphere packing problem. However, this solution of his is not yet widely accepted (see T.C. Hales [1] and W.Y. Hsiang [2]). In 1831, C.F. Gauss [1] had found that $\delta^*(B)$, the density of the densest lattice packing of spheres in R^3, is $\frac{\pi}{3\sqrt{2}}$. Since then, it has come to be widely believed that $\delta(B)$, the density of the densest packing of spheres in R^3, is $\frac{\pi}{3\sqrt{2}}$ also.

In 1900, D. Hilbert [1] listed the sphere packing problem, along with its tetrahedral analog, as part of the 18th of his famous 23 unsolved problems. In 1904, while studying Kepler's problem for lattice packings of arbitrary convex bodies, H. Minkowski [4] found some powerful criteria which can be used to determine $\delta^*(K)$ for various K. Unfortunately, in applying his method to the lattice packings of tetrahedra, Minkowski made a critical mistake. This mistake was first discovered by H. Groemer [3] about 60 years later. In 1970, D.J. Hoylman [1] sought to correct Minkowski's mistake using Minkowski's own techniques. Hoylman succeeded in proving that $\delta^*(S)=\frac{18}{49}$. However Hoylman's determination of $\mathcal{L}^\delta(S)$ had a gap in it due to an improper use of Minkowski's criteria. So far, this gap has not been mended.

For more information concerning packings, we refer the reader to C.A. Rogers [3], L. Fejes Tóth [3], and G. Fejes Tóth and W. Kuperberg [1].

The present situation with regard to the Thirteen Sphere Problem is better than that of Kepler's problem. In 1874, R. Hoppe [1] solved the

Thirteen Sphere Problem negatively. In other words, he proved that the maximal number of nonoverlapping unit spheres which can be brought into contact with a fixed unit sphere is 12. Since then, the so-called kissing numbers of general convex bodies have been studied by H. Minkowski [1], H. Hadwiger [3], H. Groemer [2], H.P.F. Swinnerton-Dyer [1], G.L. Watson [1], G.A. Kabatjanski and V.I. Levenštein [1], V.I. Levenštein [1], A.M. Odlyzko and N.J.A. Sloane [1], P.M. Gruber [2], and many others. The Swinnerton-Dyer result implies that for K any n-dimensional convex body,

$$h^*(K) \geq n(n+1).$$

In 1963, B. Grünbaum [2] conjectured that for S any n-dimensional simplex,

$$h(S) = n(n+1).$$

Recently, this conjecture was disproved by C. Zong [5]. Furthermore, when one combines this work of Zong with that of Hoylman, one obtains Phenomenon β.

The concept of blocking number is a comparatively new one in Convex and Discrete Geometry. It was first defined and studied by C. Zong [1] in 1993. It is easy to see that the blocking number is both a limiting case of and a counterpart to the kissing number. Recently, C. Zong [4] discovered Phenomenon α. In the same article, he also found an inequality between blocking numbers and *Hadwiger's covering numbers* which improved the known upper bounds on the latter.

We close this section by formulating two conjectures concerning packing densities and kissing numbers.

Conjecture 4.1. *For n sufficiently large, there exist two convex bodies K_1 and K_2 in R^n such that*

$$\delta^*(K_1) < \delta^*(K_2) \quad \textit{but} \quad \delta(K_1) > \delta(K_2).$$

Conjecture 4.1 implies another long-standing conjecture, namely the existence of a convex body K for which $\delta(K) > \delta^*(K)$. It is believed by some that such a K can be found in R^9 or R^{11}.

Conjecture 4.2. *For n sufficiently large, there exist two convex bodies K_1 and K_2 in R^n such that*

$$h^*(K_1) < h^*(K_2) \quad \textit{but} \quad h(K_1) > h(K_2).$$

Conjecture 4.2 implies the existence of a convex body K for which $h(K) > h^*(K)$. Such a K, residing in R^9, was first found by G.L. Watson [1]. Recently, C. Zong [2] has proved the existence of such bodies in R^n for any $n \geq 3$.

5

Category Phenomena

§1. Introduction

In a *metric space* $\{\Re, \rho\}$, a subset is called *meager* or of the *first category* if it can be represented as a countable union of nowhere dense subsets. We say that a property holds for most[1] elements of $\Re$ if it holds for all elements of $\Re$ that lie off a meager subset. In 1899, R. Baire [1] found that every meager subset of a *compact metric space* or a *locally compact metric space* has a dense complement. So, in a topological sense, meager sets are "small," whereas their complements are "large."

Let $\mathcal{K}$ be the family of all n-dimensional convex bodies and let δ^H be the Hausdorff metric on $\mathcal{K}$ defined by

$$\delta^H(K_1, K_2) = \max\left\{ \sup_{x \in K_1} \inf_{y \in K_2} \|x - y\|, \ \sup_{y \in K_2} \inf_{x \in K_1} \|x - y\| \right\}.$$

Blaschke's selection theorem implies that $\{\mathcal{K}, \delta^H\}$ is locally compact. Moreover, for K any convex body, $\{\partial(K), \|\cdot\|\}$ is clearly compact. Thus, *Baire's theorem* applies to both of these metric spaces.

In this chapter, we will present some strange phenomena concerning measure and category in $\{\mathcal{K}, \delta^H\}$ and $\{\partial(K), \|\cdot\|\}$. These phenomena were discovered by A.D. Aleksandrov, H. Busemann, W. Feller, P.M. Gruber, R. Schneider, and T. Zamfirescu.

---

[1]This concept is quite different from the measure-theoretic concept embodied in the phrase "almost everywhere."

§2. Gruber's Phenomenon

As usual, a convex body K is called m times differentiable if its surface is at least m times differentiable. We say that a convex body is differentiable when it is at least one times differentiable. P.M. Gruber [1] discovered the following result concerning the differentiablity of elements of $\{\mathcal{K}, \delta^H\}$.

Gruber's Phenomenon: *Most convex bodies are differentiable. However, most convex bodies are not twice differentiable.*

Proof: Let $\mathcal{A}_i$ be the family of those n-dimensional convex bodies which have a boundary point at which there exist two supporting hyperplanes making an angle $\alpha \geq \frac{1}{i}$ with each other, let $\mathcal{B}_i$ be the family of those n-dimensional convex bodies K such that for every point $x \in \partial(K)$, there exists a sphere $y + \frac{1}{i}B$ which satisfies

$$x \in \partial\Big(y + \frac{1}{i}B\Big) \subseteq K,$$

and let $\mathcal{D}_i$ be the family of all i-times differentiable n-dimensional convex bodies. We first prove

Assertion 5.1. *Concerning $\mathcal{D}_1$ and the families $\mathcal{A}_i$, we have*

$$\mathcal{K} \setminus \mathcal{D}_1 \subseteq \bigcup_{i=1}^{+\infty} \mathcal{A}_i. \tag{5.1}$$

Moreover, the families $\mathcal{A}_i$ are nowhere dense in $\{\mathcal{K}, \delta^H\}$.

First, if $K \notin \mathcal{D}_1$, then there exists a point $x \in \partial(K)$ at which two distinct supporting hyperplanes H_1 and H_2 of K can be found. Let α be the angle between H_1 and H_2. Since $\alpha > 0$, we see that

$$K \in \bigcup_{i=1}^{+\infty} \mathcal{A}_i,$$

which implies (5.1).

Second, by Lemma 4.4 it is easy to see that the families $\mathcal{A}_i$ are closed and have empty interior in $\{\mathcal{K}, \delta^H\}$. Hence, we have proved Assertion 5.1, and the first part of Gruber's phenomenon now follows.

Assertion 5.2. *Concerning $\mathcal{D}_2$ and the families $\mathcal{B}_i$, we have*

$$\mathcal{D}_2 \subseteq \bigcup_{i=1}^{+\infty} \mathcal{B}_i. \tag{5.2}$$

Moreover, the families $\mathcal{B}_i$ are nowhere dense in $\{\mathcal{K}, \delta^H\}$.

Before proving (5.2), some definitions and notation are first necessary. Assuming that $K \in \mathcal{D}_2$ and $x \in \partial(K)$, denote the *internal unit normal* of K at x by $u(x)$. Let $\|\cdot\|$ denote the Euclidean norm on R^n and ρ denote the *geodesic distance* on $\partial(B)$. Then, defining

$$\Sigma = \{\{x, y\} : \ x \in \partial(K), \ \|y\| = 1, \ \langle u(x), y\rangle = 0\}$$

and

$$\sigma(\{x_1, y_1\}, \{x_2, y_2\}) = \|x_1 - x_2\| + \rho(y_1, y_2),$$

it is easy to see that $\{\Sigma, \sigma\}$ is a compact metric space. Letting $H(x, y)$ be the 2-dimensional hyperplane determined by x, $x + u(x)$, and $x + y$, denote the *curvature* of $K \cap H(x, y)$ at x by $\varsigma\{x, y\}$. Then the assumption that $K \in \mathcal{D}_2$ implies that for a suitable integer m,

$$\max_{\{x,y\}\in\Sigma} \varsigma\{x, y\} \leq m.$$

Hence, we see that

$$K \in \mathcal{B}_m,$$

which proves (5.2).

It is well known that for every $K \in \mathcal{K}$, there is a sequence of polytopes P_1, P_2, ... such that

$$\lim_{i\to+\infty} \delta^H(K, P_i) = 0.$$

Keeping this in mind, it follows at once that the families $\mathcal{B}_i$ are closed and have empty interior in $\{\mathcal{K}, \delta^H\}$. Hence, we have proved Assertion 5.2, and the second part of Gruber's phenomenon now follows. □

§3. The Aleksandrov-Busemann-Feller Theorem

In this section, we prove a result, discovered by A.D. Aleksandrov [2] and H. Busemann and W. Feller [1], dealing with Lebesgue measure and curvature on $\{\partial(K), \|\cdot\|\}$. It is first necessary to introduce some definitions and notation. Denote the ray $\{a + \lambda(b - a) : \ \lambda \geq 0\}$ by $R(a, b)$. Let

$$A = \{p(t) : \ 0 \leq t < \alpha\}$$

be a *Jordan arc* in R^n beginning at p. Thus, the point $p(t)$ depends continuously on t, $p(t_1) \neq p(t_2)$ whenever $t_1 \neq t_2$, and $p(0) = p$. We assume that A at p has a tangent

$$\tau = \lim_{t\to 0+} R(p, p(t))$$

and that the 2-dimensional hyperplane Q_t which passes through $p(t)$ and contains τ converges as $t \to 0^+$ to a plane Q called the *osculating plane* of A at p.

Let $\varrho(t)$ denote the radius of the circle which is tangent to τ at p and that passes through p and $p(t)$. A simple argument involving similar triangles shows that

$$\varrho(t) = \frac{\|p - p(t)\|^2}{2\|p(t) - q_t\|},$$

where q_t denotes the orthogonal projection of $p(t)$ onto τ. Since

$$\|p - p(t)\|^2 = \|p - q_t\|^2 + \|p(t) - q_t\|^2$$

and

$$\lim_{t \to 0+} \frac{\|p(t) - q_t\|}{\|p - q_t\|} = 0,$$

the two limit supremums in (5.3), as well as the two limit infimums in (5.4), are equal.

Definition 5.1. *We call*

$$\varrho_s = \limsup_{t \to 0+} \varrho(t) = \limsup_{t \to 0+} \frac{\|p - q_t\|^2}{2\|p(t) - q_t\|} \tag{5.3}$$

and

$$\varrho_i = \liminf_{t \to +\infty} \varrho(t) = \liminf_{t \to 0+} \frac{\|p - q_t\|^2}{2\|p(t) - q_t\|} \tag{5.4}$$

the upper and lower radii of curvature of A at p, respectively, and call ϱ_s^{-1} and ϱ_i^{-1} the lower and upper curvatures of A at p, respectively. In addition, if $\varrho_s = \varrho_i = \varrho$, then we call ϱ the radius of curvature of A at p, and ϱ^{-1} the curvature of A at p.

Now suppose that p is a regular point of $\partial(K)$, so that there is only one supporting hyperplane H of K at p. Letting $u(p)$ be the unit normal of H which lies on the side of H containing K, suppose also that A lies on both $\partial(K)$ and some half-hyperplane of dimension 2 which has $\{p + \theta u(p) : -\infty \le \theta \le \infty\}$ as its boundary. Then the corresponding ϱ_s^{-1}, ϱ_i^{-1}, and ϱ^{-1} will be called the lower normal curvature, the upper normal curvature, and the normal curvature of $\partial(K)$ in the direction τ at p, respectively. For convenience, we denote these curvatures by r_s^{-1}, r_i^{-1}, and r^{-1}, respectively.

Definition 5.2. *Letting p be a regular point of $\partial(K)$, denote the radii of the upper and the lower normal curvatures of $\partial(K)$ in the direction τ at p by $r_s(\tau)$ and $r_i(\tau)$, respectively. Then we call*

$$J_s = \{x \in H : \|x - p\| \le \sqrt{r_s(x - p)}\}$$

and

$$J_i = \{x \in H : \|x - p\| \le \sqrt{r_i(x - p)}\}$$

the upper and the lower indicatrices of $\partial(K)$ at p, respectively. When $J_s = J_i = J$, we call J the indicatrix of $\partial(K)$ at p.

Theorem 5.1 (A.D. Aleksandrov [2], H. Busemann and W. Feller [1]). *At almost all points p of $\{\partial(K), \|\cdot\|\}$ the indicatrix exists and is star-like with p as an interior center point.*

Lemma 5.1 (Meusnier). *Assume that*

$$A = \{p(t) : \ 0 \leq t < \alpha\} \subset \partial(K),$$

that $p = p(0)$ is a regular point of $\partial(K)$, and that the acute angle between the osculating plane Q at p and $u(p)$ is β. Then

$$\varrho_s = r_s \cos\beta, \quad \varrho_i = r_i \cos\beta.$$

Proof: Let H be the supporting hyperplane of K at p, let H^* be the half-plane which contains τ and has $\{p + \theta u(p) : \ -\infty \leq \theta \leq \infty\}$ as its boundary,

$$A^* = \partial(K) \cap H^*,$$

and let τ be the positive e_1-axis. Let p_α and p^*_α denote points from the intersection of A and A^* with $x^1 = \alpha$, respectively, where $0 < \alpha < \delta$. Note that $p_\alpha \to p$ as $\alpha \to 0^+$. Then $q_\alpha = (\alpha, 0, \ldots, 0)$ is the projection of both p_α and p^*_α on τ. Hence, by (5.3) we have that

$$\varrho_s = \limsup_{\alpha \to 0+} \frac{\alpha^2}{2\|p_\alpha - q_\alpha\|}$$

and

$$r_s = \limsup_{\alpha \to 0+} \frac{\alpha^2}{2\|p^*_\alpha - q_\alpha\|},$$

with similar relations holding for ϱ_i and r_i.

According to the definition of the osculating plane, there is a plane Q_α containing τ and p_α which tends to Q as $\alpha \to 0^+$. If β_α is the acute angle between Q_α and $u(p)$, then

$$\lim_{\alpha \to 0+} \beta_\alpha = \beta \neq \frac{\pi}{2}.$$

If μ_α and μ^*_α are the coordinates of p_α and p^*_α with respect to the $u(p)$-axis, then we have

$$|\mu_\alpha| = \|p_\alpha - q_\alpha\| \cos\beta_\alpha$$

and

$$|\mu^*_\alpha| = \|p^*_\alpha - q_\alpha\|.$$

Hence, Lemma 5.1 will be established once we show that

$$\lim_{\alpha \to 0+} \frac{\mu^*_\alpha}{\mu_\alpha} = 1. \tag{5.5}$$

In what follows, we may disregard the possibility that $\mu_{\alpha(i)} = 0$ for a sequence $\alpha(i) \to 0^+$ since $\beta \neq \frac{\pi}{2}$ implies that $p_{\alpha(i)} \in \tau$ and hence that $p^*_{\alpha(i)} = p_{\alpha(i)}$ for sufficiently large i.

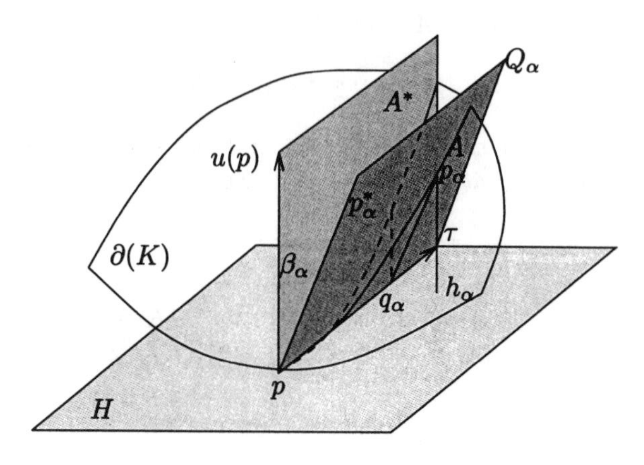

Figure 8

Letting h_α be the projection of p_α on H, we have that

$$\|h_\alpha - q_\alpha\| = \pm\mu_\alpha \tan\beta_\alpha.$$

Hence, from the hypothesis that p is a regular point of $\partial(K)$, it follows that

$$0 = \lim_{\alpha\to 0+} \frac{\mu_\alpha - \mu^*_\alpha}{\|h_\alpha - q_\alpha\|} = \lim_{\alpha\to 0+} \frac{\mu_\alpha - \mu^*_\alpha}{\mu_\alpha \tan\beta_\alpha}. \tag{5.6}$$

Since $\beta \neq \frac{\pi}{2}$ and $\mu_\alpha = \mu^*_\alpha$ whenever $h_\alpha = q_\alpha$, (5.6) implies (5.5), which completes the proof of Lemma 5.1. □

Lemma 5.2. *The lower indicatrix J_i of $\partial(K)$ at a regular point p is convex. Moreover, if p is an interior point of J_i, then both $r_s(\tau)$ and $r_i(\tau)$ at p are continuous functions of the direction τ.*

Proof: Without loss of generality, we may assume that $p = o$ is the regular point, that $x^n = 0$ is the supporting hyperplane H of K at p, and that $\partial(K)$ is locally represented about o as $x^n = f(\dot{x})$ where $\dot{x} \in H$ and $f(\dot{x}) \geq 0$ in a neighborhood of p with radius $\delta > 0$. For small $\epsilon > 0$, define

$$J_\epsilon = \left\{\dot{x} \in H : \; f(\sqrt{\epsilon}\dot{x}) \leq \frac{\epsilon}{2}, \; \dot{x} \in \frac{\delta}{\sqrt{\epsilon}} B\right\}.$$

Clearly, J_ϵ is convex. We shall show that

$$J_i = \bigcup_n \bigcap_{\epsilon < n^{-1}} J_\epsilon. \tag{5.7}$$

Setting $M_n = \bigcap_{\epsilon<n^{-1}} J_\epsilon$ and $M^n = \bigcup_{\epsilon<n^{-1}} J_\epsilon$, it is easy to check that $M_n \subseteq M_{n+1}$ and $M^n \supseteq M^{n+1}$.

If $\dot{x} \in M_n$, then for all $\epsilon < n^{-1}$ we have

$$\frac{\epsilon\|\dot{x}\|^2}{2f(\sqrt{\epsilon}\dot{x})} \geq \|\dot{x}\|^2,$$

which implies that

$$r_i(\dot{x}) = \liminf_{\epsilon\to 0+} \frac{\epsilon\|\dot{x}\|^2}{2f(\sqrt{\epsilon}\dot{x})} \geq \|\dot{x}\|^2.$$

In other words, on the one hand, we have

$$\bigcup_n M_n = \bigcup_n \bigcap_{\epsilon<n^{-1}} J_\epsilon \subseteq J_i. \tag{5.8}$$

On the other hand, if $\dot{x} \in J_i$, then

$$\frac{\epsilon\|\dot{x}\|^2}{2f(\sqrt{\epsilon}\dot{x})} \geq r_i(\dot{x}) - \eta_\epsilon(\dot{x}),$$

where $\lim_{\epsilon\to 0+} \eta_\epsilon(\dot{x}) = 0$. Since $\|\dot{x}\|^2 < r_i(\dot{x})$, we see that both $f(\sqrt{\epsilon}\dot{x}) \leq \frac{\epsilon}{2}$ and $\|\dot{x}\| \leq \delta\epsilon^{-\frac{1}{2}}$ hold for sufficiently small ϵ. Thus, we have

$$J_i \subseteq \bigcup_n \bigcap_{\epsilon<n^{-1}} J_\epsilon. \tag{5.9}$$

Clearly, (5.8) and (5.9) together imply (5.7), from which it is obvious that J_i is convex and, hence, that $r_i(\tau)$ varies continuously as a function of the direction τ.

To finish, it suffices to prove the continuity of $r_s(\tau)$. An argument similar to that establishing (5.7) establishes that

$$J_s = \bigcap_n \bigcup_{\epsilon<n^{-1}} J_\epsilon. \tag{5.10}$$

If $\dot{x} \in M^n$, then $\dot{x} \in J_\epsilon$ for some $\epsilon < n^{-1}$. Keeping the fact that $M_n \subseteq J_\epsilon$ in mind, it follows that M^n contains all segments $\dot{x}\dot{y}$ where $\dot{y} \in M_n$. In other words, for $\dot{x} \in J_s$ we have that J_s contains all segments $\dot{x}\dot{y}$ where $\dot{y} \in \bigcup_n M_n = J_i$. The continuity of $r_s(\tau)$ then follows immediately from the assumption that p is an interior point of J_i.

The proof of Lemma 5.2 is complete. □

Proof of Theorem 5.1: Denote by $v_m(X)$ the m-dimensional Lebesgue measure of a set X, and by $G(K)$ the set of regular points p of $\partial(K)$ at which the indicatrix J exists with p as an interior point and center. Theorem 5.1 will be proved by induction on the dimension n.

The base step, $n = 2$, of the induction is easy since it follows from basic Measure and Convexity Theory that

$$v_1(\partial(K) \setminus G(K)) = 0$$

for every 2-dimensional convex domain K.

Next, assume that Theorem 5.1 is true in R^{n-1}. In other words, suppose that

$$v_{n-2}(\partial(K) \setminus G(K)) = 0 \tag{5.11}$$

for every $(n-1)$-dimensional convex body K. To complete the induction we shall prove that Theorem 5.1 is also true in R^n.

Given an n-dimensional convex body K, let $\{u_1, u_2, \ldots\}$ be a countable dense subset of $\partial(B)$ and, for each i, let $\mathcal{H}(u_i)$ denote the family of all those $(n-1)$-dimensional hyperplanes which intersect K and are perpendicular to u_i. By (5.11), for every $H \in \mathcal{H}(u_i)$, we have

$$v_{n-2}(K \cap H \setminus G(K \cap H)) = 0.$$

This and Lemma 5.1 imply that

$$v_{n-1}(X_{u_i}) = 0. \tag{5.12}$$

Here, X_{u_i} is the set of the points $p \in \partial(K)$ for which the assertion

$$r_s(-\tau) = r_i(-\tau) = r_s(\tau) = r_i(\tau) > 0$$

fails to hold in some direction τ perpendicular to u_i.

Since $\{u_1, u_2, \ldots\}$ is dense in $\partial(B)$, by Lemma 5.2 we get that

$$\partial(K) \setminus G(K) = \bigcup_{i=1}^{+\infty} X_{u_i}. \tag{5.13}$$

From (5.12) and (5.13), it follows immediately that

$$v_{n-1}(\partial(K) \setminus G(K)) = 0,$$

which proves Theorem 5.1. □

§4. A Theorem of Zamfirescu

The previous section dealt with a result involving Lebesgue measure and curvature on $\{\partial(K), \|\cdot\|\}$. In this section, we return to category by proving the following theorem:

Theorem 5.2 (T. Zamfirescu [2]). *Most convex surfaces are differentiable with the property that for each point x of the surface and each vector τ tangent to the surface at x, either $r_i(\tau) = 0$ or $r_s(\tau) = +\infty$.*

Proof: For convenience, given $x \in R^n$ and two orthogonal unit vectors u and τ, set

$$H(x,u,\tau) = \{x + \alpha u + \beta \tau : \ -\infty \le \alpha \le +\infty, \ 0 \le \beta < +\infty\}.$$

When p is a regular point of $\partial(K)$, denote the supporting hyperplane of K at p by $H(p)$ and the internal unit normal of $H(p)$ at p by $u(p)$. Let $\wp$ be the family of all differentiable convex surfaces with the property that for each point p of the surface and each vector τ tangent to the surface at p, either $r_i(\tau) = 0$ or $r_s(\tau) = +\infty$. Also, let $\Im_m$ be the family of all differentiable convex surfaces which possess a point p and a tangent vector $\tau(p)$ at p such that both

$$K \cap H(p,u(p),\tau(p)) \subseteq (p + mu(p) + mB) \cap H(p,u(p),\tau(p)) \qquad (5.14)$$

and

$$\left(p + \frac{1}{m}u(p) + \frac{1}{m}B\right) \cap H(p,u(p),\tau(p)) \subseteq K \cap H(p,u(p),\tau(p)). \quad (5.15)$$

Finally, let $\Im_m^*$ be the closure of $\Im_m$ in $\{\mathcal{K}, \delta^H\}$.

On the one hand, by Assertion 5.1 and the definitions of $\wp$ and $\Im_m$, we have that

$$\wp = \mathcal{K} \setminus \bigcup_{m=1}^{+\infty} \left(\Im_m \cup \mathcal{A}_m\right).$$

On the other hand, by the selection theorems of Blaschke and Weierstraß, it is easy to see that $K \in \Im_m^*$ only if K has a boundary point p at which there is a supporting hyperplane $H(p)$ with an internal unit normal $u(p)$ and a tangent vector $\tau(p)$ perpendicular to $u(p)$ such that both (5.14) and (5.15) are satisfied. Thus, no polytope belongs to $\Im_m^*$. Since the family of polytopes is dense in $\{\mathcal{K}, \delta^H\}$, we see that $\Im_m$ is nowhere dense in $\{\mathcal{K}, \delta^H\}$.

Theorem 5.2 is proven. □

Clearly, Theorem 5.1 and Theorem 5.2 together imply the following.

Corollary 5.1. *At almost all points of most n-dimensional convex surfaces,*

$$r_i = r_s = +\infty.$$

§5. The Schneider-Zamfirescu Phenomenon

Although complicated to prove, Theorems 5.1 and 5.2 are both fundamental and believable. In stark contrast to these however, R. Schneider [2] and T. Zamfirescu [1] found the following strange result.

The Schneider-Zamfirescu Phenomenon: *At most points of most convex surfaces, both $r_i(\tau) = 0$ and $r_s(\tau) = +\infty$ for every tangent vector τ.*

To verify this phenomenon, we will need some definitions and notation in additon to those already introduced in the previous sections. Given $p \in R^n$, two orthogonal unit vectors u and τ, and $0 \leq r < r' \leq +\infty$, define

$$\begin{aligned} M = M_{p,u,\tau,r,r'} &= \{x \in H(p,u,\tau) \cap H(p,\tau,u) : \ \|x - (p + ru)\| \geq r, \\ &\qquad \|x - (p + r'u)\| \leq r'\}, \\ U_p &= \{p + \mu u : \ -\infty \leq \mu \leq +\infty\}, \end{aligned}$$

and

$$\alpha(M) = \arctan r' - \arctan r.$$

We call M a *sickle*. Moreover, we say that this sickle is adapted to K at p if $p \in \partial(K)$ and u is an internal normal to K at p. The corresponding r and r' are called the radii of the sickle. Finally, for i, j, k, $l \in \mathbb{Z}$, let $\mathcal{E}_{i,j,k,l}$ denote the family of all those convex bodies $K \in \mathcal{K}$ satisfying the following condition:

** *There exists a point $x \in \partial(K)$ such that for each point $p \in \partial(K) \cap (x + \frac{1}{i}\ \mathrm{int}\ (B))$, there is a sickle M adapted to K at p for which*

$$\alpha(M) = \frac{\pi}{2} - \frac{1}{j}, \tag{5.16}$$

$$\partial(K) \cap H(p, u(p), \tau(p)) \cap \left(p + \frac{1}{k}\ \mathrm{int}(B)\right) \subseteq M, \tag{5.17}$$

and

$$U_p \text{ contains a point } y \text{ with } y + \frac{1}{l}\ \mathrm{int}(B) \subseteq K. \tag{5.18}$$

Lemma 5.3 (R. Schneider [2]). *$\mathcal{E}_{i,j,k,l}$ is closed in $\{\mathcal{K}, \delta^H\}$.*

Proof: Let K_m, $m \in \mathbb{Z}$, be a sequence in $\mathcal{E}_{i,j,k,l}$ which converges to some convex body $K \in \mathcal{K}$. Then, for each m we may choose a point $x_m \in \partial(K_m)$ for which the condition ** is satisfied. Without loss of generality, we may assume that the sequence x_m, $m \in \mathbb{Z}$, converges to a point $x \in \partial(K)$. Let $p \in \partial(K) \cap (x + \frac{1}{i}\ \mathrm{int}(B))$; then p is the limit of a sequence $p_m \in \partial(K_m)$, $m \in \mathbb{Z}$. Since $\|x - p\| < \frac{1}{i}$, we have $\|x_m - p_m\| < \frac{1}{i}$ for sufficiently large m. For each of these m, there exists a sickle M_m at p_m such that (5.16), (5.17), and (5.18) hold with corresponding K_m, u_m, τ_m, r_m, r'_m, and y_m. Without loss of generality, we may assume that

$$\lim_{m \to +\infty} u_m = u, \qquad \lim_{m \to +\infty} \tau_m = \tau,$$

$$\lim_{m\to+\infty} r_m = r, \qquad \lim_{m\to+\infty} r'_m = r', \quad \text{and} \quad \lim_{m\to+\infty} y_m = y.$$

Here, r' can be infinite. Whether it is infinite or not, we have $r < r'$, and not merely $r \leq r'$, because

$$\arctan r'_m - \arctan r_m = \alpha(M_m) = \frac{\pi}{2} - \frac{1}{j}.$$

Denoting the sickle $M_{p,u,\tau,r,r'}$ more briefly by M, we have $\alpha(M) = \frac{\pi}{2} - \frac{1}{j}$. The vector u is an internal normal vector to K at p, and so M is adapted to K at p. The point y lies on the line U_p. Since $y_m + \frac{1}{l}\ \text{int}(B) \subseteq K_m$ for sufficiently large m, it follows that $y + \frac{1}{l}\ \text{int}(B) \subseteq K$. We assert that

$$\lim_{m\to+\infty} K_m \cap H(p_m, u_m, \tau_m) = K \cap H(p, u, \tau). \tag{5.19}$$

By way of proof, on the one hand, assume that $v \in K \cap H(p, u, \tau)$. Let v_m be the point in $K_m \cap H(p_m, u_m, \tau_m)$ nearest to v and suppose that the sequence v_m, $m \in \mathbb{Z}$, does not converge to v. Then there exists a positive number ϕ such that

$$K_m \cap H(p_m, u_m, \tau_m) \cap (v + \phi\ \text{int}(B)) = \emptyset$$

for infinitely many $m \in \mathbb{Z}$. For each of these m, there exists a hyperplane which separates K_m and $H(p_m, u_m, \tau_m) \cap (v + \phi\ \text{int}(B))$. A subsequence of these hyperplanes converges to a hyperplane which separates K and $H(p, u, \tau) \cap (v + \phi\ \text{int}(B))$. Since

$$v \in K \cap H(p, u, \tau) \cap (v + \phi\ \text{int}(B)),$$

it follows that v lies in this hyperplane and that the hyperplane separates K and $H(p, u, \tau)$. But this is a contradiction since U_p meets an interior point of K. Hence, the sequence v_m, $m \in \mathbb{Z}$, converges to v. On the other hand, if $v_m \in K_m \cap H(p_m, u_m, \tau_m)$ and $\lim_{m\to+\infty} v_m = v$, then, clearly,

$$v \in K \cap H(p, u, \tau).$$

This proves (5.19).

Next, we turn to prove that

$$\partial(K) \cap H(p, u, \tau) \cap \left(p + \frac{1}{k}\ \text{int}(B)\right) \subseteq M. \tag{5.20}$$

Letting $w \in \partial(K) \cap H(p, u, \tau) \cap (p + \frac{1}{k}\ \text{int}(B))$, by (5.19) we see that w is the limit of a sequence $w_m \in \partial(K_m) \cap H(p_m, u_m, \tau_m)$, $m \in \mathbb{Z}$. Since $\|w - p\| < \frac{1}{k}$, we have that $\|w_m - p_m\| < \frac{1}{k}$ for sufficiently large m. By (5.17), for M_m, we have that $w_m \in M_m$ for each of these m; that is

$$\begin{cases} w_m \in H(p_m, u_m, \tau_m) \cap H(p_m, \tau_m, u_m), \\ \|w_m - (p_m + r_m u_m)\| \geq r_m, \\ \|w_m - (p_m + r'_m u_m)\| \leq r'_m, \end{cases}$$

which implies that

$$\begin{cases} w \in H(p,u,\tau) \cap H(p,\tau,u), \\ \|w-(p+ru)\| \geq r, \\ \|w-(p+r'u)\| \leq r'. \end{cases}$$

Therefore, $w \in M$, and so (5.20) is established.

In summary, we have shown that the sickle M, which is adapted to K at p, satisfies (5.16), (5.17), and (5.18). Since p is an arbitrary point of $\partial(K) \cap (x + \frac{1}{i}\ \mathrm{int}(B))$, the convex body K is an element of $\mathcal{E}_{i,j,k,l}$. This completes the proof of Lemma 5.3. □

Lemma 5.4 (R. Schneider [2]). *$\mathcal{E}_{i,j,k,l}$ is nowhere dense in $\{\mathcal{K}, \delta^H\}$.*

Proof: Since $\mathcal{E}_{i,j,k,l}$ is closed in $\{\mathcal{K}, \delta^H\}$, it is sufficient to show that for any $K \in \mathcal{K}$ and $\epsilon > 0$, there exists a convex body $K' \in \mathcal{K} \setminus \mathcal{E}_{i,j,k,l}$ such that

$$\delta^H(K, K') < 2\epsilon.$$

First, we construct an auxiliary convex curve in the plane. Set

$$\Gamma_1 = \{(\xi, \eta) \in R^2 : \ \eta = \xi^3, \ \xi \geq 0\}$$

and

$$\Gamma_2 = \{(\xi, \eta) \in R^2 : \ \eta = \xi^{\frac{3}{2}}, \ \xi \geq 0\}.$$

Let $0 < \alpha < 1$ and define $a_1 = (\alpha, \alpha^3)$. Assuming that a_m has been defined, define a_{m+1} to be the unique point on Γ_1 with a smaller ξ-coordinate than a_m for which the closed segment $a_{m+1}a_m$ is tangent to Γ_2. Then

$$\Gamma = \{(0,0)\} \cup \bigcup_{m=1}^{+\infty} a_{m+1}a_m$$

is a convex curve having right lower and right upper curvatures of 0 and $+\infty$, respectively at $(0,0)$. Let $f_\alpha : \ [0,\alpha] \mapsto \mathbb{R}$ be the function whose graph is Γ.

Next, for $x = (x^1, x^2, \ldots, x^n) \in R^n$, set $\|x\|^* = \left(\sum_{i=1}^{n-1} (x^i)^2\right)^{\frac{1}{2}}$ and let

$$K_\alpha = \{(x^1, x^2, \ldots, x^n) : \ \|x\|^* \leq \alpha, \ f_\alpha(\|x\|^*) \leq x^n \leq \alpha^3\}.$$

Clearly, K_α is a convex body of revolution with Γ as (part of) its *meridian curve.* The set $\{(x^1, x^2, \ldots, x^n) \in K_\alpha : \ x^n = \alpha^3\}$ is called the base of K_α, and the point $(0, 0, \ldots, 0)$ is called the vertex of K_α. We shall use similar terminology for congruent copies of K_α.

Now let $K \in \mathcal{K}$, $\epsilon > 0$, and $i \in \mathbb{Z}$ be given. Then there exists a convex polytope P such that

$$\delta^H(K, P) < \epsilon \tag{5.21}$$

and such that the sphere $x + \frac{1}{2i}\,\mathrm{int}(B)$ contains a vertex of P for each $x \in \partial(P)$. At each vertex v_g of P, choose a supporting hyperplane H_g of P such that $H_g \cap P = \{v_g\}$. For $\delta > 0$, let $H_g(\delta)$ be the hyperplane parallel to H_g at a distance δ which is on the same side of H_g as P. Let $P_g(\delta)$ denote the polytope which is cut off from P by $H_g(\delta)$. Assume that δ has been chosen small enough so that $P_g(\delta) \cap P_h(\delta) = \emptyset$ whenever $g \neq h$, the diameter of each $P_g(\delta)$ is less than $\frac{1}{2i}$, and the polytope

$$P_\delta = \overline{P \setminus \bigcup_g P_g(\delta)}$$

(here, $\overline{X}$ indicates the closure of X) satisfies

$$\delta^H(P, P_\delta) < \epsilon. \tag{5.22}$$

Continuing, choose α so small that each $P_g(\delta)$ contains some congruent copy $K_{\alpha,g}$ of K_α whose base is a subset of $H_g(\delta)$. Taking

$$K' = \mathrm{conv}\left\{P_\delta \cup \bigcup_g K_{\alpha,g}\right\},$$

we have that $P_\delta \subseteq K' \subseteq P$. Therefore, by (5.21) and (5.22) we obtain

$$\delta^H(K, K') < 2\epsilon.$$

Finally, we prove that $K' \notin \mathcal{E}_{i,j,k,l}$. It is easy to see that for each g, the vertex p_g of $K_{\alpha,g}$ has a neighborhood relative to the boundary of $K_{\alpha,g}$ that is contained in the boundary of K'. Therefore, the lower and upper curvatures of $\partial(K')$ at p_g are 0 and $+\infty$, respectively, for every vector tangent at p_g to K'. This means that there does not exist a sickle M adapted to K' at p_g for which $\alpha(M) < \frac{\pi}{2}$ and $\partial(K') \cap H(p_g, u_g, \tau_g) \cap (p_g + \phi\,\mathrm{int}(B)) \subseteq M$ for some positive ϕ. Moreover, by the way that K' was constructed, we see that for every point $x \in \partial(K')$ there exists a vertex p_g of K' such that

$$p_g \in x + \frac{1}{i}\,\mathrm{int}(B).$$

This shows that $K' \notin \mathcal{E}_{i,j,k,l}$ for any j, k, $l \in \mathbb{Z}$ and, hence, proves Lemma 5.4. □

Verification of the Schneider-Zamfirescu Phenomenon: Let $\mathcal{G}$ denote the family of all nondifferentiable n-dimensional convex bodies and $\mathcal{F}$ denote the family of all n-dimensional differentiable convex bodies for which the boundary points at which both $r_i(\tau) = 0$ and $r_s(\tau) = +\infty$ for every tangent vector τ form a dense subset of $\{\partial(K), \|\cdot\|\}$. We split the verification up into two assertions.

Assertion 5.3 (R. Schneider [2]).

$$\mathcal{K} \setminus \mathcal{F} = \mathcal{G} \cup \bigcup_{i,j,k,l \in \mathbb{Z}} \mathcal{E}_{i,j,k,l}.$$

On the one hand, from the definitions of $\mathcal{F}$ and $\mathcal{E}_{i,j,k,l}$, it is easy to see that $\mathcal{E}_{i,j,k,l} \subseteq \mathcal{K} \setminus \mathcal{F}$. On the other hand, if $K \in \mathcal{K} \setminus \mathcal{F}$ and K is differentiable, then there exists a nonempty, relatively open subset X of $\partial(K)$ such that at each point $p \in X$, there exists a tangent vector τ at p for which either $r_s(\tau) < +\infty$ or $r_i(\tau) > 0$. By the definitions of $r_i(\tau)$ and $r_s(\tau)$, this implies that there exists a sickle M adapted to K at p with $\alpha(M) < \frac{\pi}{2}$ and a positive number ϕ such that $\partial(K) \cap H(p, u, \tau) \cap (p + \phi \operatorname{int}(B)) \subseteq M$. Since K is smooth, the line U_p meets the interior of K. Thus, we have

$$X \subseteq \bigcup_{j,k,l \in \mathbb{Z}} X_{j,k,l}, \tag{5.23}$$

where $X_{j,k,l}$ denotes the set of all points $x \in \partial(K)$ at which there exists a sickle M adapted to K for which (5.16), (5.17), and (5.18) hold.

Repeating an argument from the proof of Lemma 5.3, we notice that each $X_{j,k,l}$ is closed. Thus, by (5.23) we can find three indices j_0, k_0, and l_0 such that X_{j_0,k_0,l_0} has a nonempty interior. In consequence, there exists a point $p \in \partial(K)$ and an index i such that

$$\partial(K) \cap \left(p + \frac{1}{i} \operatorname{int}(B)\right) \subseteq X_{j_0,k_0,l_0}.$$

This means that $K \in \mathcal{E}_{i,j_0,k_0,l_0}$, and so Assertion 5.3 is proven.

Assertion 5.4 (T. Zamfirescu [1]). *If $K \in \mathcal{F}$, then for most points p of $\{\partial(K), \|\cdot\|\}$, $r_i(\tau) = 0$ and $r_s(\tau) = +\infty$ for every tangent vector τ at p.*

Let Y denote the set of points $p \in \partial(K)$ such that both $r_i(\tau) = 0$ and $r_s(\tau) = +\infty$ for every tangent vector τ at p. For a fixed positive number γ, define

$$B_{p,\gamma,\tau} = \{x \in H(p, u(p), \tau) : \ \|x - p - \gamma u(p)\| = \gamma\}.$$

Let N_m be the set of points $p \in \partial(K)$ such that for every tangent vector τ at p,

$$\operatorname{card}\left\{\partial(K) \cap B_{p,\gamma,\tau} \cap \left(p + \frac{1}{m} B\right)\right\} \geq 2.$$

We claim that $\partial(K) \setminus N_m$ is nowhere dense in $\{\partial(K), \|\cdot\|\}$.

Let Z be an open set in $\{\partial(K), \|\cdot\|\}$ and take $p_0 \in Z \cap Y$. Since $p_0 \in Y$, for each tangent vector τ_0 at p_0, there exist two points $x_i(\tau_0)$ and $x_s(\tau_0) \in B_{p_0,\gamma,\tau_0}$, the first in the interior and the second in the exterior of K, at distances $d_i(\tau_0)$ and $d_s(\tau_0) < \frac{1}{m}$ from p_0. Since K is differentiable, there exists an open set $Z'_{p_0,\tau_0} \subseteq Z$ containing p_0 and an open set $Z''_{p_o,\tau_0} \subseteq \partial(B)$ containing τ_0 such that for every point $x \in Z'_{p_0,\tau_0}$ and every vector $\tau \in Z''_{p_0,\tau_0}$ which is tangent to K at x, the two points $x_i(\tau)$, $x_s(\tau) \in B_{x,\gamma,\tau}$ at distances $d_i(\tau_0)$ and $d_s(\tau_0)$ from x still lie in the interior and exterior of K, respectively. The sets Z''_{p_0,τ_0} obtained in this way for all vectors τ_0 tangent to K at p_0 form an open covering of $\{\tau : \langle \tau, u(p_0)\rangle = 0,\ \tau \in \partial(B)\}$. Choose a finite subcovering $Z''_{p_0,\tau_1}, \dots, Z''_{p_0,\tau_q}$ and consider the corresponding open sets $Z'_{p_0,\tau_1}, \dots, Z'_{p_0,\tau_q}$ on $\partial(K)$. Furthermore, let Z''' be an open set containing p_0 such that

$$\tau \in \bigcup_{i=1}^{q} Z''_{p_0,\tau_i}$$

for every vector τ tangent to K at some point $x \in Z'''$. Take

$$Z^* = Z''' \cap \bigcap_{i=1}^{q} Z'_{p_0,\tau_i}.$$

For each $x \in Z^*$, every vector tangent to K at x lies in some Z''_{p_0,τ_i} and x belongs to the corresponding Z'_{p_0,τ_i}. Thus, for each $x \in Z^*$ and every vector τ tangent to K at x, the circle $B_{x,\gamma,\tau}$ has two points distinct from x at a distance less than m^{-1} from x, one in the interior and the other in the exterior of K. Therefore, $Z^* \subseteq N_m$. Keeping in mind the assumption that Y is dense in $\{\partial(K), \|\cdot\|\}$, we have established our claim that $\partial(K) \setminus N_m$ is nowhere dense in $\{\partial(K), \|\cdot\|\}$. This immediately implies that

$$\bigcup_{m=1}^{+\infty} \left(\partial(K) \setminus N_m\right)$$

is a meager set in $\{\partial(K), \|\cdot\|\}$.

Hence, for most points p in $\{\partial(K), \|\cdot\|\}$, we have that for every vector τ tangent to K at p and every natural number m, there exists a point $p' \neq p$ on $\partial(K) \cap B_{p,\gamma,\tau}$ at a distance at most m^{-1} from p. In consequence, there exists a sequence of points on $\partial(K) \cap B_{p,\gamma,\tau}$ converging to p. This clearly implies that γ lies between $r_i(\tau)$ and $r_s(\tau)$. Now let γ take the values 2, $\frac{1}{2}$, 3, $\frac{1}{3}$, ... successively. Each time, for most points p of $\partial(K)$ and every tangent vector τ at p, we have

$$r_i(\tau) \leq \gamma \leq r_s(\tau).$$

Thus, for most points p of $\{\partial(K), \|\cdot\|\}$ and every tangent vector τ at p,

$$r_i(\tau) = 0 \quad \text{and} \quad r_s(\tau) = +\infty.$$

This proves Assertion 5.4.

Gruber's phenomemon, Lemma 5.4, and Assertion 5.3 imply that $\mathcal{K}\setminus\mathcal{F}$ is meager. Assertion 5.4 then immediately implies the Schneider-Zamfirescu phenomenon. □

§6. Some Remarks

K. Reidemeister [1] was perhaps the first to study the connection between convexity and differentiablity. Then came the well-known discovery of A.D. Aleksandrov [2] and H. Busemann and W. Feller [1] which was the subject of Section 3. Although the proof of the theorem of Aleksandrov, Busemann, and Feller is deep and complicated, the assertion itself is rather believable.

In 1959, V. Klee [1] proved that most convex bodies in $\{\mathcal{K}, \delta^H\}$ are differentiable in the sense of catagory. Then P.M. Gruber [1] found the unexpected phenomenon bearing his name which was discussed in Section 2.

V. Klima and I. Netuka [1] sharpened Gruber's discovery by applying *Hölder's condition* to show that most convex bodies in $\{\mathcal{K}, \delta^H\}$ are not of class $C_{1+\alpha}$ for any $\alpha > 0$ at "many" boundary points. From one point of view, the Schneider-Zamfirescu phenomenon is clearly a generalization of Gruber's phenomenon. From another point of view, the Schneider-Zamfirescu phenomenon is a counterpart and contrast to the theorem of Aleksandrov, Busemann, and Feller.

Besides the results discussed in this chapter, there are many other results of Convex and Discrete Geometry involving category. We mention only three such results here. In 1982, T. Zamfirescu [3] showed that no *geodesic arcs* pass through most boundary points of most convex bodies. In 1986, P.M. Gruber [2] found that most convex bodies in $\{\mathcal{K}, \delta^H\}$ have densest lattice packings whose "kissing number" is less than or equal to $2n^2$. (Note that this upper bound is a counterpart to the lower bound obtained in Lemma 4.7.) In 1988, P.M. Gruber [3] proved that most convex bodies in $\{\mathcal{K}, \delta^H\}$ meet the boundary of their *minimal circumscribed ellipsoids* at exactly $\frac{1}{2}n(n+3)$ points.

For further results concerning convexity and catagory, we refer the reader to P.M. Gruber [4] and T. Zamfirescu [4].

6

The Busemann-Petty Problem

§1. Introduction

The search for relationships between a convex body and its projections or sections has a long history. In 1841, A. Cauchy found that the surface area of a convex body can be expressed in terms of the areas of its projections as follows:

$$s(K) = \frac{1}{\omega_{n-1}} \int_{\partial(B)} \bar{v}(P_u(K)) d\lambda(u).$$

Here, $s(K)$ denotes the surface area of a convex body $K \subset R^n$, $\bar{v}(X)$ denotes the $(n-1)$-dimensional "area" of a set $X \subset R^{n-1}$, P_u denotes the orthogonal projection from R^n to the hyperplane $H_u = \{x \in R^n : \langle x, u \rangle = 0\}$ determined by a unit vector u of R^n, and λ denotes surface-area measure on $\partial(B)$. In contrast, the closely related problem of finding an expression for the volume of K in terms of the areas of its projections $P_u(K)$ (or the areas of its sections $I_u(K) = K \cap H_u$) proved to be unexpectedly and extremely difficult.

More than 100 years after Cauchy's discovery, H. Busemann [1] proved the following result:

Theorem 6.1′. *For every convex body $K \subset R^n$ $(n \geq 3)$,*

$$v(K)^{n-1} \geq \frac{\omega_n^{n-2}}{n\omega_{n-1}^n} \int_{\partial(B)} \bar{v}\left(I_u(K)\right)^n d\lambda(u),$$

with equality being attained only if K is an ellipsoid centered at the origin o.

Clearly, *Busemann's theorem* has the following consequence:

Corollary 6.1. *Let E be an ellipsoid with center o and let K be an arbitrary convex body. If $\bar{v}(I_u(E)) < \bar{v}(I_u(K))$ for every unit vector u, then $v(E) < v(K)$.*

The following simple-sounding problem, raised by H. Busemann and C.M. Petty [1] in 1956, is a generalization of this fascinating corollary.

The Busemann-Petty Problem: *Let C_1 and C_2 be centrally symmetric convex bodies in R^n. Does $\bar{v}(I_u(C_1)) < \bar{v}(I_u(C_2))$ for all u imply that $v(C_1) < v(C_2)$?*

One might expect an affirmative answer to this problem when the hypothesis of convexity is replaced by that of being star-like with respect to the origin or when the hypothesis of symmetry is dropped altogether. This is not the case: In 1960, H. Busemann [3] found that both convexity and symmetry are necessary. Since then, this problem has attracted the interest of many mathematicians. In 1975, D. Larman and C.A. Rogers [1] showed that the problem has a negative answer for dimensions $n \geq 12$. The analogous problem concerning projections was formulated by G.C. Shephard [1] in 1964 and solved by R. Schneider [1] in 1967. This chapter will deal with some of these exciting results connected with the Busemann-Petty problem.

§2. Steiner Symmetrization

Recall that $\mathcal{K}$ is the metric space of all convex bodies in R^n equipped with the Hausdorff metric. Given u a unit vector in R^n, let U_u be the line in R^n through the origin with direction vector u and let H_u be the hyperplane $\{x \in R^n : \langle x, u\rangle = 0\}$. For K a convex body in R^n, let $\aleph_u(K)$ denote the family of segments $\{X : X = K \cap (U_u + x),\ x \in H_u\}$ and let $M_u(K)$ denote the set of the centers of the segments $X \in \aleph_u(K)$.

Definition 6.1: *Given u a unit vector, the associated Steiner symmetrization S_u is a transformation of $\mathcal{K}$ into itself defined as follows: Given $K \in \mathcal{K}$, $S_u(K)$ is the union of all those segments obtained by translating each of the segments $X \in \aleph_u(K)$ parallel to the line U_u so that the center of X is contained in H_u.*

Steiner symmetrization is one of the most important techniques in Convex and Discrete Geometry. The following fundamental lemma is necessary for the proof of Busemann's theorem in the next section.

Lemma 6.1 (J. Steiner [1]). *Denote the family of all those convex bodies which can be obtained from a given convex body K through the application* *of a finite sequence of Steiner symmetrizations by $\mathcal{A}$. Let r be the positive number for which $v(rB) = v(K)$. Then there exists a sequence of convex bodies $K_1, K_2, \ldots \in \mathcal{A}$ such that*

$$\lim_{i\to+\infty} \delta^H(K_i, rB) = 0.$$

Proof: Define

$$\gamma(K) = \max_{x\in K}\{\|x - o\|\}.$$

Since Steiner symmetrization preserves $v(K)$ and does not increase $\gamma(K)$, we see that $r \le \gamma(A) \le \gamma(K)$ for every $A \in \mathcal{A}$. Hence, setting

$$\beta = \min_{A\in\mathcal{A}}\{\gamma(A)\},$$

by Blaschke's selection theorem we can find a sequence of convex bodies K_1, $K_2, \ldots \in \mathcal{A}$ and a convex body Q satisfying $\gamma(Q) = \beta$ and $v(Q) = v(K)$ such that

$$\lim_{i\to+\infty} \delta^H(K_i, Q) = 0.$$

To finish, we must show that $Q = rB$. So suppose that $Q \ne rB$. Since $v(Q) = v(K) = v(rB)$ and both Q and rB are compact and convex, a small ball $z + \epsilon B$ can be found such that

$$z + \epsilon B \subset rB \setminus Q.$$

Let G be the cap on $\partial(rB)$ which is obtained by radially projecting $z+\epsilon B$ from o onto $\partial(rB)$, let G_u be the cap on $\partial(rB)$ which is the reflection of G about the plane H_u, and let Y_u be the cone with vertex o and base G_u. An elementary argument shows that

$$\gamma\left(S_u(Q) \cap Y_u\right) < \beta. \tag{6.1}$$

Thus, if $u_1, u_2, \ldots, u_m$ are chosen so that the congruent caps G_{u_1}, G_{u_2}, $\ldots$, G_{u_m} cover $\partial(rB)$, then (6.1) implies that

$$\gamma\left(S_{u_1}(S_{u_2}\ldots S_{u_m}(Q)\ldots)\right) < \beta,$$

which contradicts the way that β was defined. Therefore, $Q = rB$ must be satisfied. □

§3. A Theorem of Busemann

Theorem 6.1 (H. Busemann [1]). *Let $K_1, K_2, \ldots, K_{n-1}$ be convex bodies in R^n $(n \ge 3)$. Then*

$$\prod_{i=1}^{n-1} v(K_i) \ge \frac{\omega_n^{n-2}}{n\omega_{n-1}^n} \int_{\partial(B)} \prod_{i=1}^{n-1} \bar{v}\left(I_u(K_i)\right)^{\frac{n}{n-1}}\, d\lambda(u),$$

with equality being attained only when all the K_i are homothetic ellipsoids centered at the origin o.

Clearly, Theorem 6.1′ follows from Theorem 6.1 by setting K_1, K_2, ... and K_{n-1} all equal to K. The following characterization of ellipsoids is necessary for the complicated proof of Theorem 6.1.

Lemma 6.2. *Consider a convex body K. If $M_u(K)$ is a coplanar set of points for each unit vector u, then K is an ellipsoid.*

Proof: First, we show that among those ellipsoids which contain K there is a unique one of minimal volume.[1] Without loss of generality, we may assume that $rB \subseteq K \subseteq r'B$. Denote by $\mathcal{E}$ the family of all those ellipsoids which contain K and set

$$v = \min_{E \in \mathcal{E}} \{v(E)\}.$$

Then we can find a sequence of ellipsoids $E_1, E_2, \ldots$ such that $K \subseteq E_i \subseteq r'B$ and

$$v = \lim_{i \to +\infty} v(E_i).$$

Let o_i be the center of E_i and $z_j(E_i)$, $j = 1, 2, \ldots, n$, be the n axes of E_i. Then $(o_i, z_1(E_i), \ldots, z_n(E_i))$ can be understood as a point of the cube $\{(x^1, \ldots, x^{n(n+1)}) : |x^l| \leq 2r'\}$. It thus follows from Weierstraß's selection theorem that there exists an ellipsoid E containing K and with volume v.

The existence of E disposed of, we now turn to prove its uniqueness. Let E' be another ellipsoid containing K and with volume v. Without loss of generality, by an appropriate translation followed by appropriate shears, we may assume that $E = B$, and so $v = v(E) = \omega_n$. Without loss of generality, by an appropriate rotation we may assume that E' is described by the inequality

$$\sum_{i=1}^{n} \frac{(x^i - t^i)^2}{r_i^2} \leq 1.$$

Since $v(E') = v = \omega_n$, it must be that $\prod_{i=1}^{n} r_i = 1$. The convex body $\frac{1}{2}(E + E')$ is an ellipsoid described by the inequality

$$\sum_{i=1}^{n} \frac{1}{2}\left(1 + \frac{1}{r_i^2}\right)\left(x^i - \frac{t^i}{1 + r_i^2}\right)^2 \leq 1 - \frac{1}{2}\sum_{i=1}^{n} \frac{(t^i)^2}{1 + r_i^2}.$$

Furthermore, $K \subseteq E \cap E' \subseteq \frac{1}{2}(E + E')$ and

[1] This ellipsoid is usually called the minimal circumscribed ellipsoid or the *Loewner ellipsoid* of K.

$$v\left(\frac{1}{2}(E+E')\right) = \omega_n \prod_{i=1}^{n} \left\{ \left(\frac{1}{2}(1+r_i^{-2})\right)^{-\frac{1}{2}} \left(1 - \frac{1}{2}\sum_{i=1}^{n} \frac{(t^i)^2}{1+r_i^2}\right)^{\frac{1}{2}} \right\}$$

$$\leq \omega_n \prod_{i=1}^{n} \left(\frac{1}{2}(1+r_i^{-2})\right)^{-\frac{1}{2}} \tag{6.2}$$

$$\leq \omega_n \prod_{i=1}^{n} r_i^{\frac{1}{2}} = \omega_n = v.$$

By the way v was defined, we see that $v(\frac{1}{2}(E+E'))$ must be v. Then (6.2) implies that each $r_i = 1$ and each $t^i = 0$. Hence, $E' = E$.

Having proved the existence and uniqueness of E, we now proceed to prove that $K = E$ when K is as in the statement of the lemma.

Obviously, the center o of E is an interior point of K. Define a bijection $f : \partial(K) \mapsto \partial(E)$ by setting $f(x)$ equal to the unique intersection point of $\overrightarrow{ox}$ with $\partial(E)$. If at each point $x \in \partial(K)$ there exists a unique supporting hyperplane of K and this hyperplane is parallel to that of E at $f(x)$, then it follows from some elementary Differential Geometry and the minimality of the volume of E that $K = E$. Otherwise, an elementary geometric argument produces a line U_u and a corresponding hyperplane H_u such that $M_u(K)$ and $M_u(E)$ lie in two different hyperplanes, H^K and H^E, respectively. We shall obtain a contradiction in this second situation.

i. *H^K and H^E are parallel.*

Then $H^K = H^E + \delta u$ for some $\delta \neq 0$, and so from the definition of $M_u(K)$ we get

$$K \subseteq E + \delta u,$$

which contradicts the uniqueness of E.

ii. *H^K and H^E are not parallel.*

Assuming that $o \in H^K \cap H^E$, clearly, R^n can be expressed as $U_u \oplus (H^K \cap H^E) \oplus L$ where L is a line in H^E with unit direction vector e_n. This means that every point x of R^n can be uniquely expressed as

$$x = x^1 u + y + x^n e_n,$$

where $y \in H^K \cap H^E$. Thus, there exists a constant α such that

$$y + x^n e_n + \alpha x^n u \in H^K$$

whenever $y + x^n e_n \in H^E$. Hence, defining a linear transformation T by

$$x^1 u + y + x^n e_n \to (x^1 + \alpha x^n)u + y + x^n e_n,$$

we see that $T(E)$ is an ellipsoid, $K \subseteq T(E)$, and $v(T(E)) = v$, which contradicts the uniqueness of E.

In conclusion, $K = E$, which proves our lemma. □

We next obtain an analytical expression for $\prod_{i=1}^{n-1} v(K_i)$.

Lemma 6.3. *Denote by dx_i the $(n-1)$-dimensional volume element on the hyperplane H_u to which $I_u(K_i)$ belongs and, by $S(o, x_1, \ldots, x_{n-1})$ (or more briefly by S) the simplex with vertices o, x_1, ..., and x_{n-1}. Then*

$$\prod_{i=1}^{n-1} v(K_i) = \frac{(n-1)!}{2} \int_{\partial(B)} \int_{I_u(K_1)} \cdots \int_{I_u(K_{n-1})} \bar{v}(S) dx_{n-1} \ldots dx_1 d\lambda(u). \tag{6.3}$$

Proof: Considering each K_i to be a subset of R_i^n, set

$$R^{n(n-1)} = R_1^n \oplus R_2^n \oplus \cdots \oplus R_{n-1}^n$$

and

$$K = K_1 \oplus K_2 \oplus \cdots \oplus K_{n-1}.$$

For convenience, write the n coordinates of R_i^n as $x_i^1, x_i^2, \ldots, x_i^n$. Then we clearly have

$$\prod_{i=1}^{n-1} v(K_i) = v(K) = \int_K dx_1^1 \ldots dx_1^n \ldots dx_{n-1}^1 \ldots dx_{n-1}^n. \tag{6.4}$$

Defining $\nu_1, \nu_2, \ldots$, and ν_{n-1} by means of the system of linear equations

$$x_i^n = \sum_{j=1}^{n-1} \nu_j x_i^j, \quad i = 1, \ldots, n-1,$$

we may introduce new coordinates $x_1^1, \ldots, x_1^{n-1}$, $\nu_1, \ldots, x_{n-1}^1, \ldots, x_{n-1}^{n-1}$, ν_{n-1} in $R^{n(n-1)}$. It is easy to see that the *Jacobian* of this coordinate transformation is

$$J = \left| x_i^j \right|_{i,\ j=1,\ldots,n-1.} \tag{6.5}$$

(The set where $J = 0$ is too small to affect the integral on the right of (6.3) and off this set our coordinate transformation is well defined.)

Taking $\mu = \left(1 + \sum_{j=1}^{n-1} \nu_j^2\right)^{-\frac{1}{2}}$, the unit normal u of the hyperplane $x^n = \nu_1 x^1 + \cdots + \nu_{n-1} x^{n-1}$ may be specified by either

$$u^j = \mu \nu_j, \quad j = 1, \ldots, n-1; \ u^n = -\mu$$

or

$$u^j = -\mu \nu_j, \quad j = 1, \ldots, n-1; \ u^n = \mu.$$

Letting θ be the angle between u and the x^n-axis, we have that $\mu = |\cos\theta|$, and so the area element of $\partial(B)$ at u is $d\lambda(u) = \mu^{-1} du^1 \ldots du^{n-1}$. Here, for integration purposes we may disregard the small set of those planes which are parallel to the x^n-axis. Some technical calculations show that

$$\left|\frac{\partial u^j}{\partial \nu_i}\right|_{i,\, j=1,\ldots,n-1} = \mu^{3(n-1)} \begin{vmatrix} \mu^{-2}-\nu_1^2 & -\nu_1\nu_2 & \ldots & -\nu_1\nu_{n-1} \\ -\nu_2\nu_1 & \mu^{-2}-\nu_2^2 & \ldots & -\nu_2\nu_{n-1} \\ \vdots & \vdots & \ddots & \vdots \\ -\nu_{n-1}\nu_1 & -\nu_{n-1}\nu_2 & \ldots & \mu^{-2}-\nu_{n-1}^2 \end{vmatrix}$$

$$= \mu^{3(n-1)}\left(\mu^{-2(n-1)} - \mu^{-2(n-2)}\sum_{j=1}^{n-1}\nu_j^2\right)$$

$$= \mu^{n+1},$$

and so

$$\begin{aligned} d\lambda(u) &= \mu^{-1}du^1\ldots du^{n-1} = \mu^n d\nu_1\ldots d\nu_{n-1} \\ &= |\cos^n\theta|\, d\nu_1\ldots d\nu_{n-1}. \end{aligned} \tag{6.6}$$

In addition, the volume element dx_j of the hyperplane $x_j^n = \nu_1 x_j^1 + \cdots + \nu_{n-1}x_j^{n-1}$ is

$$dx_j = |\sec\theta|dx_j^1\ldots dx_j^{n-1}. \tag{6.7}$$

Now, if we interpret the points $x_1, x_2, \ldots, x_{n-1}$ as lying in the same space R^n, then (6.5) shows that $|J|/(n-1)!$ is the volume of the projection of the simplex S with vertices o, $x_1, \ldots, x_{n-1}$ onto the plane $x^n = 0$. Since these points lie in the hyperplane $\langle x, u\rangle = 0$, we have

$$\bar{v}(S)|\cos\theta| = \frac{|J|}{(n-1)!}$$

and hence

$$|J\sec\theta| = (n-1)!\bar{v}(S). \tag{6.8}$$

Keeping in mind (6.5), (6.6), (6.7), (6.8), and the fact that every $I_u(K_i)$ is counted twice while integrating over $\partial(B)$, it follows from (6.4) that

$$\begin{aligned} \prod_{i=1}^{n-1} v(K_i) &= \int_K |J|dx_1^1\ldots dx_1^{n-1}\ldots dx_{n-1}^1\ldots dx_{n-1}^{n-1}d\nu_1\ldots d\nu_{n-1} \\ &= \frac{1}{2}\int_{\partial(B)}\int_{I_u(K_1)}\cdots\int_{I_u(K_{n-1})} |J\sec\theta|dx_{n-1}\ldots dx_1 d\lambda(u) \\ &= \frac{(n-1)!}{2}\int_{\partial(B)}\int_{I_u(K_1)}\cdots\int_{I_u(K_{n-1})} \bar{v}(S)dx_{n-1}\ldots dx_1 d\lambda(u), \end{aligned}$$

which proves Lemma 6.3. □

We now turn to the last ingredient in the proof of Busemann's theorem.

Lemma 6.4. *Let $K_1, K_2, \dots, K_m$ be convex bodies in R^m. Then*

$$\int_{K_1} \cdots \int_{K_m} v(S) dx_1 \dots dx_m \geq \frac{2\omega_{m+1}^{m-1}}{(m+1)!\omega_m^{m+1}} \prod_{i=1}^{m} v(K_i)^{\frac{m+1}{m}}, \tag{6.9}$$

with equality being attained only when all the K_i are homothetic ellipsoids centered at the origin o.

Proof: The key idea of the proof is to apply various Steiner symmetrizations efficiently to $K_1, K_2, \dots, K_m$ while showing that in the process the integral on the left-hand side of (6.9) does not increase. For convenience, we abbreviate the integral in (6.9) to $I(K_1, \dots, K_m)$ and the determinant $\left|a_i^j\right|_{i,j=1,\dots,m}$ to $\left\lfloor a_i^j \right\rfloor$. Without loss of generality, take H_u to be the hyperplane $x^m = 0$.

For $x_i \in K_i$, denote the reflection of x_i through the center of $K_i \cap (U_u + x_i)$ by x_i^*, and the image of x_i in $S_u(K_i)$ by $\overline{x_i}$. Writing $x_i = (x_i^1, x_i^2, \dots, x_i^m)$, some easy geometric arguments show that

$$v(S(o, x_1, \dots x_m)) = \pm \frac{1}{m!} \left\lfloor x_i^j \right\rfloor,$$

$$v(S(o, x_1^*, \dots, x_m^*)) = \pm \frac{1}{m!} \left\lfloor x_i^{*j} \right\rfloor,$$

and

$$\left\lfloor \overline{x_i^j} \right\rfloor = - \left\lfloor \overline{x_i^{*j}} \right\rfloor.$$

Hence,

$$2v(S(o, \overline{x_1}, \dots, \overline{x_m})) = 2v(S(o, \overline{x_1^*}, \dots, \overline{x_m^*})) = \frac{1}{m!} \left| \left\lfloor \overline{x_i^j} \right\rfloor - \left\lfloor \overline{x_i^{*j}} \right\rfloor \right|.$$

Moreover, we clearly have

$$x_i^m - x_i^{*m} = \overline{x_i^m} - \overline{x_i^{*m}}$$

and

$$x_i^j = x_i^{*j} = \overline{x_i^j} = \overline{x_i^{*j}} \quad \text{for } 1 \leq j \leq m-1.$$

Therefore, we get

$$\left\lfloor \overline{x_i^j} \right\rfloor - \left\lfloor \overline{x_i^{*j}} \right\rfloor = \left\lfloor x_i^j \right\rfloor - \left\lfloor x_i^{*j} \right\rfloor$$

and

$$\begin{aligned} v(S(o, x_1, \dots, x_m)) &+ v(S(o, x_1^*, \dots, x_m^*)) \\ &= \frac{1}{m!} \left(\left| \left\lfloor x_i^j \right\rfloor \right| + \left| \left\lfloor x_i^{*j} \right\rfloor \right| \right) \geq \frac{1}{m!} \left(\left| \left\lfloor x_i^j \right\rfloor - \left\lfloor x_i^{*j} \right\rfloor \right| \right) \\ &= \frac{1}{m!} \left(\left| \left\lfloor \overline{x_i^j} \right\rfloor - \left\lfloor \overline{x_i^{*j}} \right\rfloor \right| \right) \\ &= 2v(S(o, \overline{x_1}, \dots, \overline{x_m})). \end{aligned} \tag{6.10}$$

Since

$$I(K_1,\ldots,K_m) = \int_{K_1}\cdots\int_{K_m} v(S(o,x_1,\ldots,x_m))dx_m\ldots dx_1$$
$$= \int_{K_1}\cdots\int_{K_m} v(S(o,x_1^*,\ldots,x_m^*))dx_m^*\ldots dx_1^*$$

nd

$$dx_i = dx_i^* = d\overline{x_i} = d\overline{x_i^*},$$

rom (6.10) it follows that

$(K_1,\ldots,K_m)$

$$= \frac{1}{2}\Big(\int_{K_1}\cdots\int_{K_m} v(S(o,x_1,\ldots x_m))dx_m\ldots dx_1$$
$$+\int_{K_1}\cdots\int_{K_m} v(S(o,x_1^*,\ldots,x_m^*))dx_m^*\ldots dx_1^*\Big)$$
$$= \int_{K_1}\cdots\int_{K_m}\frac{1}{2}\Big(v(S(o,x_1,\ldots,x_m))+v(S(o,x_1^*,\ldots,x_m^*))\Big)dx_m\ldots dx_1$$
$$\geq \int_{K_1}\cdots\int_{K_m} v(S(o,\overline{x_1},\ldots,\overline{x_m}))dx_m\ldots dx_1$$
$$= \int_{S_u(K_1)}\cdots\int_{S_u(K_m)} v(S(o,\overline{x_1},\ldots,\overline{x_m}))d\overline{x_m}\ldots d\overline{x_1}$$
$$= I(S_u(K_1),\ldots,S_u(K_m)). \tag{6.11}$$

From (6.11) and Lemma 6.1, it follows that

$$I(K_1,\ldots,K_m)\geq I(r_1B,\ldots,r_mB), \tag{6.12}$$

here B is the m-dimensional unit ball and each $r_i = \left(\frac{v(K_i)}{\omega_m}\right)^{\frac{1}{m}}$.

Now suppose that equality is attained in (6.11). Then we must have quality in (6.10) for any selection of points $x_i \in K_i$, in particular for any election of points $x_i \in M_u(K_i)$. For such points, $x_i = x_i^*$ and $\overline{x_i} = \overline{x_i^*} \in H_u$, o the right-hand side of (6.10) vanishes, and, thus, the points $o, x_1,\ldots,x_m$ re coplanar in R^m. We conclude that equality is attained in (6.11) only hen all the sets $M_u(K_i)$ lie in a single plane.

From this and Lemma 6.2, we see that equality is attained in (6.12) only hen the K_i are homothetic ellipsoids centered at the origin o.

Finally, applying (6.3) with $n = m+1$ and each $K_i = r_iB^m$ (here, the pper index of B indicates its dimension), we obtain

$$\prod_{i=1}^m v(r_iB^{m+1}) = \frac{m!}{2}\int_{\partial(B^{m+1})} I(r_1B^m,\ldots,r_mB^m)d\lambda(u).$$

Consequently,

$$I(K_1,\dots,K_m) \geq I(r_1B^m,\dots,r_mB^m)$$
$$= \frac{2\omega_{m+1}^{m-1}}{(m+1)!\omega_m^{m+1}} \prod_{i=1}^{m} v(K_i)^{\frac{m+1}{m}},$$

with equality being attained only when the K_i are homothetic ellipsoids centered at the origin o.

Lemma 6.4 is proven. □

Clearly, Theorem 6.1 is an immediate consequence of Lemmas 6.3 and 6.4.

§4. The Larman-Rogers Phenomenon

In working on the Busemann-Petty problem, D. Larman and C.A. Rogers [1] found the following result.

The Larman-Rogers Phenomenon: *When $n \geq 12$, there are two centrally symmetric convex bodies C_1, $C_2 \subset R^n$ such that $v(C_1) > v(C_2)$, but for all unit vectors u,*

$$\bar{v}(I_u(C_1)) < \bar{v}(I_u(C_2)).$$

The proof that Larman and Rogers gave of their result is very complicated. The proof that will be presented here, due to A.A. Giannopoulos [1], is constructive and shows the following: *For α and β positive, set*

$$C(\alpha,\beta) = \left\{(x^1,\dots,x^n) : \sum_{i=1}^{n-1}(x^i)^2 \leq \alpha^2, \ \ |x^n| \leq \beta\right\}.$$

Then, for $n \geq 8$, suitable α and β can be found so that $v(C(\alpha,\beta)) = v(B)$ but for all unit vectors u,

$$\bar{v}(I_u(C(\alpha,\beta))) < \bar{v}(I_u(B)).$$

For convenience in what follows, we shall use the following abbreviation

$$I_k = \int_0^1 (1-t^2)^{\frac{k-1}{2}}\,dt = \int_0^{\frac{\pi}{2}} \cos^k\theta\, d\theta = \frac{\omega_k}{2\omega_{k-1}}. \tag{6.13}$$

Lemma 6.5. *Take* $m = \left(\frac{\alpha}{\beta}\right)^2$ *and set*

$$f(x) = \begin{cases} 1, & x = 0 \\ \frac{\sqrt{1+mx^2}}{x} \int_0^x (1-t^2)^{\frac{n-2}{2}}\, dt, & 0 < x \le 1. \end{cases}$$

Then

$$\sup_{u \in \partial(B)} \bar{v}(I_u(C(\alpha,\beta))) = 2\omega_{n-2}\alpha^{n-2}\beta \sup_{0 \le x \le 1} f(x).$$

Proof: Assuming that $u = (u^1, \ldots, u^n)$ with $u^n > 0$, set $\varphi = \operatorname{arccot}\left(\frac{\alpha}{\beta}\right)$ and $\psi_u = \arccos u^n$. Obviously, $\bar{v}(I_u(C(\alpha,\beta)))$ depends only on u^n. We now consider the following two cases:

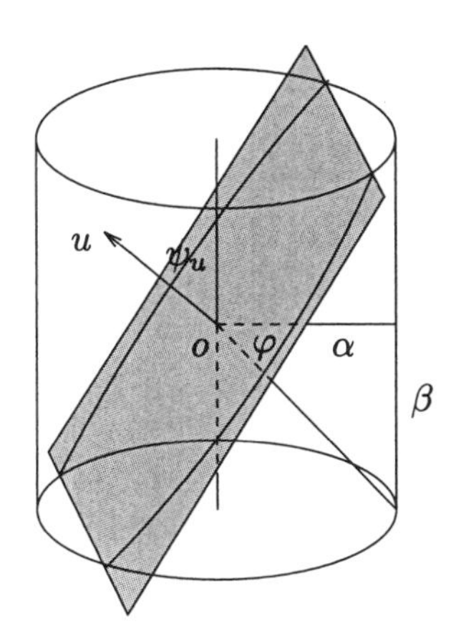

Figure 9

Case 1. $0 \le \psi_u \le \varphi$.

Then $u^n = \cos\psi_u \ge \cos\varphi = \frac{\alpha}{\sqrt{\alpha^2+\beta^2}}$. Let P be the orthogonal projection onto the hyperplane $x^n = 0$. It is easy to see that in this case

$$P(I_u(C(\alpha,\beta))) = \left\{ (x^1, \ldots, x^n) : \ x^n = 0, \ \sum_{i=1}^{n-1} (x^i)^2 \le \alpha^2 \right\}.$$

Consequently, by (6.13) we have

$$\begin{aligned} \bar{v}(I_u(C(\alpha,\beta))) &= \frac{\bar{v}(P(I_u(C(\alpha,\beta))))}{u^n} \\ &= \frac{\omega_{n-1}\alpha^{n-1}}{u^n} \\ &\le \omega_{n-1}\alpha^{n-2}\sqrt{\alpha^2+\beta^2} \\ &= 2\omega_{n-2}\alpha^{n-2}\beta f(1). \end{aligned} \tag{6.14}$$

Case 2. $\varphi \le \psi_u \le \frac{\pi}{2}$.

Then $\cot^2 \psi_u \le \left(\frac{\alpha}{\beta}\right)^2$. Set

$$u^* = \frac{1}{\sqrt{1-(u^n)^2}}(u^1, \ldots, u^{n-1}, 0)$$

and let P^* be the orthogonal projection onto the hyperplane $\langle x, u^* \rangle = 0$. If $x \in I_u(C(\alpha, \beta))$, letting $t = x^n$ so that $|t| \le \beta$, and $\mu = \left\{\sum_{i=1}^{n-1}(x^i)^2\right\}^{1/2}$ so that $\mu \le \alpha$, we have

$$P^*(x) = x - \langle x, u^* \rangle u^* = x + (t \cot \psi_u) u^*$$

and

$$\|P^*(x) - te_n\|^2 = \mu^2 - t^2 \cot^2 \psi_u.$$

It is also clear that

$$P^*(x) - te_n \in \{u^*, e_n\}^\perp,$$

where $\{x_1, \ldots, x_i\}^\perp$ denotes the subspace of R^n which is perpendicular to all the vectors $x_1, \ldots, x_i$.

Conversely, if $\langle y, u^* \rangle = 0$, $y^n = t$, and

$$\|y - te_n\|^2 = \mu^2 - t^2 \cot^2 \psi_u,$$

then

$$z = y - (t \cot \psi_u) u^* \in I_u(C(\alpha, \beta)),$$

$$\sum_{i=1}^{n-1} (z^i)^2 = \mu^2,$$

$z^n = t$, and $P^*(z) = y$. Hence, $P^*(I_u(C(\alpha, \beta))) \cap \{x : x^n = t\}$ is an $(n-2)$-dimensional ball lying in $te_n + \{u^*, e_n\}^\perp$ with radius $\sqrt{\alpha^2 - t^2 \cot^2 \psi_u}$ and center te_n.

Thus,

$$\bar{v}(P^*(I_u(C(\alpha, \beta)))) = \omega_{n-2} \int_{-\beta}^{\beta} \left(\alpha^2 - t^2 \cot^2 \psi_u\right)^{\frac{n-2}{2}} dt,$$

and so

$$\begin{aligned} \bar{v}(I_u(C(\alpha, \beta))) &= \frac{\omega_{n-2}}{|\langle u, u^* \rangle|} \int_{-\beta}^{\beta} \left(\alpha^2 - t^2 \cot^2 \psi_u\right)^{\frac{n-2}{2}} dt \\ &= \frac{\omega_{n-2}}{\sin \psi_u} \int_{-\beta}^{\beta} \left(\alpha^2 - t^2 \cot^2 \psi_u\right)^{\frac{n-2}{2}} dt \qquad (6.15) \\ &= \omega_{n-2} \sqrt{1 + \cot^2 \psi_u} \int_{-\beta}^{\beta} \left(\alpha^2 - t^2 \cot^2 \psi_u\right)^{\frac{n-2}{2}} dt. \end{aligned}$$

Setting $x = \frac{\beta}{\alpha}\cot\psi_u$ and then changing variables by letting $w = \frac{xt}{\beta}$, we see that (6.15) becomes

$$\begin{aligned}\bar{v}(I_u(C(\alpha,\beta))) &= \omega_{n-2}\alpha^{n-2}\sqrt{1+mx^2}\int_{-\beta}^{\beta}\left(1-\frac{x^2t^2}{\beta^2}\right)^{\frac{n-2}{2}}dt\\ &= \omega_{n-2}\alpha^{n-2}\beta\frac{\sqrt{1+mx^2}}{x}\int_{-x}^{x}(1-w^2)^{\frac{n-2}{2}}dw\\ &= 2\omega_{n-2}\alpha^{n-2}\beta f(x).\end{aligned}\tag{6.16}$$

Upon observing that $\varphi \le \psi_u \le \frac{\pi}{2}$ is equivalent to $0 \le x \le 1$, Lemma 6.5 follows from (6.14) and (6.16). □

Lemma 6.6. *For $n \ge 3$, the function $f(x)$ defined in Lemma 6.5 is decreasing on the interval $[0,1]$ if and only if $m \le \frac{n-2}{3}$.*

Proof: By differentiating, we see that $f(x)$ is decreasing if and only if

$$x(1+mx^2)(1-x^2)^{\frac{n-2}{2}} \le \int_0^x (1-t^2)^{\frac{n-2}{2}}dt, \quad 0 \le x \le 1. \tag{6.17}$$

For

$$g(x) = x(1+mx^2)(1-x^2)^{\frac{n-2}{2}} - \int_0^x (1-t^2)^{\frac{n-2}{2}}dt,$$

one can verify that $g'(x) \le 0$ if and only if

$$3m(1-x^2) \le (n-2)(1+mx^2).$$

Hence, since $g(0) = 0$, we see that (6.17) holds if and only if $m \le \frac{n-2}{3}$, and so Lemma 6.6 is proven. □

Lemma 6.7 (K. Ball [2]). *Set $r_n = \left(\frac{I_n}{I_{n-1}}\right)^{n-1}$ and $s_n = 2\sqrt{n}I_n$. Then r_n is decreasing but s_n is increasing.*

Proof: By some elementary analysis, we see that

$$\log\left(1+\frac{1}{x}\right) < \frac{2x+1}{2x(x+1)}, \quad x > 0 \tag{6.18}$$

and

$$\log\left(1+\frac{1}{x-1}\right) > \frac{2x+1}{2x^2+1}, \quad x > 1. \tag{6.19}$$

Differentiating and applying (6.18) and (6.19), we find that the function $f(x) = (\frac{x+1}{x})^{2x+1}$ is decreasing on $(0,+\infty)$, whereas the function $g(x) = (\frac{x}{x-1})^{2x-1-\frac{1}{x}}$ is increasing on $(1,+\infty)$. These facts imply that for $n \ge 3$,

$$\left(\frac{n-1}{n-2}\right)^{\frac{n-3}{2n-5}} \left(\frac{n+1}{n}\right)^{\frac{n}{2n-1}} < \frac{n}{n-1}. \tag{6.20}$$

Since $\frac{I_{n-1}}{I_{n+1}} = \frac{n+1}{n}$, set

$$h(n) = \left(\frac{r_n}{r_{n+1}}\right)^{\frac{1}{2n-1}} = \frac{I_n}{I_{n-1}} \left(\frac{n+1}{n}\right)^{\frac{n}{2n-1}};$$

(6.20) implies that $h(n) < h(n-2)$ for $n \geq 3$. Moreover, it can be easily checked that $\lim_{n\to+\infty} h(n) = 1$. Hence, $h(n) \geq 1$ for all $n \geq 1$, i.e., the sequence r_n is decreasing.

Similarly, taking

$$h^*(n) = \frac{s_{n+1}}{s_n},$$

we see that $\lim_{n\to+\infty} h^*(n) = 1$ and that

$$\frac{h^*(n+2)}{h^*(n)} = \sqrt{\frac{n(n+2)}{(n+1)(n+3)}\,\frac{n+2}{n+1}} < 1.$$

Therefore, $h^*(n) > 1$, which implies that the sequence s_n is increasing.

Lemma 6.7 is proven. □

We now turn to the last ingredient in the verification of the Larman-Rogers phenomenon.

Lemma 6.8. *For $n \geq 8$, we can choose α and β such that*

$$v(C(\alpha,\beta)) = \omega_n, \tag{6.21}$$

$$2\omega_{n-2}\alpha^{n-2}\beta f(0) < \omega_{n-1}, \tag{6.22}$$

and

$$m \leq \frac{n-2}{3}. \tag{6.23}$$

Proof: Since $v(C(\alpha,\beta)) = 2\omega_{n-1}\alpha^{n-1}\beta$, (6.21) is true if and only if

$$\alpha^{n-1}\beta = I_n. \tag{6.24}$$

In view of (6.24) and the fact that $f(0) = 1$, (6.22) is satisfied if

$$\frac{2\omega_{n-2}I_n}{\alpha} < \omega_{n-1};$$

in other words, if

$$\alpha > \frac{I_n}{I_{n-1}}. \tag{6.25}$$

Setting $\alpha^{n-1}\beta = I_n$ and $\alpha = I_n I_{n-1}^{-1}$, we get

$$m = \left(\frac{\alpha}{\beta}\right)^2 = \left(\frac{\alpha^n}{I_n}\right)^2 = \left\{\left(\frac{I_n}{I_{n-1}}\right)^{n-1}\frac{1}{I_{n-1}}\right\}^2 = \left(\frac{r_n}{I_{n-1}}\right)^2.$$

Thus, (6.23) will be satisfied if

$$r_n < \sqrt{\frac{n-2}{3}}I_{n-1} = \frac{1}{2\sqrt{3}}\sqrt{\frac{n-2}{n-1}}s_{n-1}. \tag{6.26}$$

We only need to verify (6.26) for $n = 8$ since Lemma 6.7 shows that the sequence r_n is decreasing and the sequence s_n is increasing. In fact, we have

$$r_8 = \left(\frac{I_8}{I_7}\right)^7 = 0.64637\ldots$$

and

$$\sqrt{\frac{8-2}{3}}I_7 = 0.64649\ldots .$$

Hence, choosing α slightly larger than $I_n I_{n-1}^{-1}$ and taking $\beta = I_n\alpha^{1-n}$, it follows from (6.24), (6.25), and (6.26) that (6.21), (6.22), and (6.23) hold simultaneously.

Lemma 6.8 is proven. □

Taking $C_1 = C(\alpha, \beta)$ and $C_2 = B$, the Larman-Rogers phenomenon follows immediately from Lemmas 6.5, 6.6, and 6.8.

§5. Schneider's Phenomenon

Let $\mathcal{C}$ denote the family of all n-dimensional centrally symmetric convex bodies and let $\mathcal{M}$ denote the family of all n-dimensional centrally symmetric convex bodies that can be arbitrarily closely approximated, in the sense of the Hausdorff metric, by Minkowski sums of finitely many segments. In 1967, while confirming a conjecture of G.C. Shephard, R. Schneider [1] found

Schneider's Phenomenon: *Suppose $C \in \mathcal{C}$, $D \in \mathcal{M}$, and for every unit vector u,*

$$\bar{v}(P_u(D)) > \bar{v}(P_u(C)).$$

Then

$$v(D) > v(C).$$

If $C_1 \in \mathcal{C}\backslash\mathcal{M}$ is a convex body whose supporting function h_{C_1} is at least $n+4$ times differentiable and whose curvature function is everywhere positive, then there exists a $C_2 \in \mathcal{C}$ such that for every unit vector u,

$$\bar{v}(P_u(C_2)) > \bar{v}(P_u(C_1)),$$

but

$$v(C_2) < v(C_1).$$

The verification of Schneider's phenomenon requires the following lemmas concerning *spherical harmonics* and Minkowski sums of finitely many segments.

Lemma 6.9. *Let $k \geq n+2$ be an even integer. For every function $G(u)$ which is $k+1$ times differentiable on $\partial(B)$ and satisfies $G(-u) = G(u)$, there exists a continuous function $g(u)$ defined on $\partial(B)$ and satisfying $g(-u) = g(u)$ such that*

$$G(u) = \int_{\partial(B)} |\langle u, w\rangle| g(w) d\lambda(w).$$

Proof: To begin, we need to state some results about spherical harmonics. For proofs, we refer the reader to C. Müller [1] or the appendix of R. Schneider [3].

Assertion 6.1. *There are exactly*

$$N(n,m) = \frac{(2m+n-2)\Gamma(m+n-2)}{\Gamma(m+1)\Gamma(n-1)} \tag{6.27}$$

n-dimensional orthogonal linearly independent spherical harmonics $Y_{m,j}$ of order $m \geq 1$ ($N(n,0) = 1$). Moreover, every complete system $Y_{m,j}$ of n-dimensional orthogonal linearly independent spherical harmonics of order m satisfies

$$\sum_{j=1}^{N(n,m)} Y_{m,j}^2 = \frac{N(n,m)}{n\omega_n}. \tag{6.28}$$

Assertion 6.2. *For any n-dimensional spherical harmonic Y_m of order m,*

$$\int_{\partial(B)} |\langle u, w\rangle| Y_m(w) d\lambda(w) = \lambda_m Y_m(u), \tag{6.29}$$

where

$$\begin{aligned} \lambda_m &= \frac{\pi^{\frac{n-1}{2}}}{2^{m-1}\Gamma\left(m+\frac{n-1}{2}\right)} \int_{-1}^{1} |t| \frac{d^m}{dt^m}(1-t^2)^{m+\frac{n-3}{2}} dt \\ &= \frac{(-1)^{\frac{m-2}{2}} \pi^{\frac{n-1}{2}} \Gamma(m-1)}{2^{m-2}\Gamma\left(\frac{m}{2}\right)\Gamma\left(\frac{m+n-1}{2}\right)} \end{aligned} \tag{6.30}$$

when m is even and $\lambda_m = 0$ when m is odd.

Assertion 6.3. *Given $F(u)$ a continuous function on $\partial(B)$, define*

$$\alpha_{m,j} = \int_{\partial(B)} F(w) Y_{m,j}(w) d\lambda(w), \tag{6.31}$$

$$a_m = \left(\sum_{j=1}^{N(n,m)} \alpha_{m,j}^2 \right)^{\frac{1}{2}}, \tag{6.32}$$

and

$$Y_m(u) = \sum_{j=1}^{N(n,m)} \alpha_{m,j} Y_{m,j}(u). \tag{6.33}$$

Then

$$\int_{\partial(B)} |F(w)|^2 d\lambda(w) = \sum_{m=0}^{+\infty} a_m^2 \tag{6.34}$$

and

$$F(u) = \lim_{r \to 1^-} \sum_{m=0}^{+\infty} r^m Y_m(u). \tag{6.35}$$

Assertion 6.4. *Let Δ denote the Laplace operator on $\partial(B)$. Then*

$$\Delta Y_m(u) + m(m+n-2) Y_m(u) = 0. \tag{6.36}$$

With these preliminaries done, we set $F(u) = G(u)$ in Assertion 6.3. Then to prove the lemma, by Assertion 6.2 it suffices to show that the convergence of the series in the following two equations is uniform:

$$G(u) = \sum_{m=0}^{+\infty} Y_m(u), \tag{6.37}$$

$$g(u) = \sum_{2|m,\ m \geq 0} \lambda_m^{-1} Y_m(u). \tag{6.38}$$

First, by (6.30) and *Stirling's formula* we see that for m even,

$$|\lambda_m^{-1}| = O\left(m^{\frac{n+2}{2}}\right). \tag{6.39}$$

Second, in (6.31) and (6.32) replace $F(u)$ by $\Delta^{\frac{k}{2}} G(u)$ and denote the resulting $\alpha_{m,j}$ and a_m by $\beta_{m,j}$ and b_m, respectively. Since $G(u)$ is $k+1$ times differentiable, it follows from (6.34) that

$$\sum_{m=0}^{+\infty} b_m^2 < +\infty,$$

and so $\lim_{m\to+\infty} b_m = 0$. Applying (6.33), (6.31), (6.32), *Green's integral formula*, (6.36), and the *Cauchy-Schwarz inequality*, we get

$$\begin{aligned} a_m^2 &= \int_{\partial(B)} G(w)Y_m(w)d\lambda(w) \\ &= \left(\frac{-1}{m(m+n-2)}\right)^{\frac{k}{2}} \int_{\partial(B)} Y_m(w)\Delta^{\frac{k}{2}}G(w)d\lambda(w) \\ &\le m^{-k} \sum_{j=1}^{N(n,m)} \alpha_{m,j} \int_{\partial(B)} Y_{m,j}(w)\Delta^{\frac{k}{2}}G(w)d\lambda(w) \\ &\le m^{-k}a_m b_m. \end{aligned}$$

Thus,

$$a_m = o\left(m^{-k}\right). \tag{6.40}$$

Then, by (6.33), the Cauchy-Schwarz inequality, (6.28), (6.27), Stirling's formula, (6.40), and (6.39), we obtain

$$\begin{aligned} |Y_m(u)| &= \left|\sum_{j=1}^{N(n,m)} \alpha_{m,j}Y_{m,j}(u)\right| \le a_m \left(\sum_{j=1}^{N(n,m)} Y_{m,j}^2(u)\right)^{\frac{1}{2}} \\ &= a_m \left(\frac{N(n,m)}{n\omega_n}\right)^{\frac{1}{2}} = O\left(m^{\frac{n-2}{2}}\right)a_m = o\left(m^{\frac{n-2}{2}-k}\right) \end{aligned}$$

and

$$\left|\lambda_m^{-1}Y_m(u)\right| = o\left(m^{n-k}\right).$$

These estimates imply that the convergence in both (6.37) and (6.38) is uniform, and so Lemma 6.9 is proven. □

Lemma 6.10. *For every integer $k > 0$, there exists an n-dimensional centrally symmetric convex body $C \notin \mathcal{M}$ whose supporting function $h_C(u)$ is k times differentiable on $\partial(B)$ and whose curvature function $\psi_C(u)$ is everywhere positive on $\partial(B)$.*

Proof: We must first prove a basic assertion due to H. Minkowski [3]:

Assertion 6.5. *For any given n-dimensional convex body K and any given positive number ϵ, there exists an analytic function $\Omega(x)$ defined on R^n such that*

$$K^* = \{x \in R^n : \ \Omega(x) \le 1\} \tag{6.41}$$

is a convex body whose curvature function $\psi_{K^*}(u)$ is positive and whose Hausdorff distance $\delta^H(K^*, K)$ to K is less than ϵ.

It is easy to see that there exists a polytope P for which

$$K \subseteq P \subseteq K + \frac{\epsilon}{2}B.$$

Assume that $u_1, u_2, \ldots, u_N$ are the external unit normals of the facets F_1, $F_2, \ldots, F_N$ of P. Then the hyperplane which contains F_j has the equation

$$L_j(x) = \frac{1}{h_P(u_j)} \sum_{i=1}^{n} x^i u_j^i - 1 = 0,$$

and so we have

$$P = \{x \in R^n : \; L_j(x) \leq 0, \; 1 \leq j \leq N\}. \tag{6.42}$$

Taking

$$M > \max\left\{\log N, \frac{2\log N}{\epsilon} \max_{1 \leq j \leq N}\{h_P(u_j)\}\right\},$$

we may, and do, define an analytic function

$$\Omega(x) = \frac{1}{N}\sum_{i=1}^{N} e^{ML_i(x)}.$$

Then, by (6.41) and (6.42) we get

$$K \subseteq P \subseteq K^*. \tag{6.43}$$

Moreover, from the definitions of P and u_j, we see that if $x \notin P + \frac{\epsilon}{2}B$, then for at least one index j,

$$L_j(x) \geq \frac{\epsilon}{2h_P(u_j)}.$$

Thus, for any $x \notin P + \frac{\epsilon}{2}B$, we have

$$\Omega(x) \geq \frac{1}{N}\exp\left(\frac{M\epsilon}{2\max_{1 \leq j \leq N}\{h_P(u_j)\}}\right) > 1.$$

This implies that

$$K^* \subseteq P + \frac{\epsilon}{2}B \subseteq K + \epsilon B. \tag{6.44}$$

We next show that K^* is convex with a curvature function $\psi_{K^*}(u)$ that is positive on $\partial(B)$. Since $ze^z \geq -e^{-1}$ always, $M > \log N > \log 3$, and $L_j(x) \geq 0$ for at least one j whenever $x \in \partial(K^*)$, we see that

$$\begin{aligned} e^M \sum_{i=1}^{n} x^i \frac{\partial \Omega}{\partial x^i} &= \frac{1}{N}\sum_{j=1}^{N} M(L_j(x)+1)e^{M(L_j(x)+1)} \\ &\geq \frac{1}{N}\left(Me^M - \frac{N-1}{e}\right) \neq 0. \end{aligned}$$

This implies that there is a supporting hyperplane of K^* at any $x \in \partial(K^*)$. Differentiating Ω twice along an arbitrary direction, we obtain

$$\frac{d^2\Omega}{dt^2} = \frac{M^2}{N} \sum_{j=1}^{N} e^{ML_j(x)} \left(\frac{dL_j(x)}{dt} \right)^2 > 0, \tag{6.45}$$

which means that K^* is convex and the curvature function $\psi_{K^*}(u)$ is positive on $\partial(B)$.

Assertion 6.5 clearly follows from (6.43), (6.44), and (6.45).

Note that $h_{K^*}(u)$ is analytic on $\partial(B)$ for the convex body K^* just constructed.

We next prove that the regular octahedron O cannot be arbitrarily closely approximated by Minkowski sums of finitely many segments. We will do this by deducing a contradiction from the assumption that for any $\epsilon_i > 0$, there exist segments $\eta_{i,1}u_{i,1}, \eta_{i,2}u_{i,2}, \ldots, \eta_{i,l}u_{i,l}$ such that

$$\delta^H \left(O, \sum_{j=1}^{l} \eta_{i,j}u_{i,j} \right) < \epsilon_i.$$

(Here, each $u_{i,j}$ is a unit direction vector and each $\eta_{i,j} > 0$ is a length.)

Let $\delta > 0$ be much smaller than any n-dimensional solid angle of O and let $T_1, T_2, \ldots, T_w$ be a tiling of $\partial(B)$ such that for each i, the solid angle formed by the unit vectors whose endpoints belong to T_i is smaller than δ. Clearly, then each

$$P_{i,j} = \sum_{u_{i,k} \in T_j} \eta_{i,k}u_{i,k}$$

is a polytope which is not homothetic to O. Letting ϵ_i be a sequence of positive numbers with limit 0 and applying Blaschke's selection theorem to each of the families $\mathcal{P}_j = \{P_{i,j} : i = 1, 2, \ldots\}$, $j = 1, 2, \ldots, w$, in turn, we see that

$$O = \sum_{j=1}^{w} C_j,$$

where each C_j is a convex body that is not homothetic to O. However, one can easily verify that O is indecomposable, and so we have obtained the expected contradiction.

In conclusion, from Assertion 6.5 and from what we have just shown about O, it follows that the family of all those n-dimensional centrally symmetric convex bodies C for which h_C is k times differentiable and ψ_C is everywhere positive on $\partial(B)$ is dense in $\mathcal{C}$ and that the family of all those n-dimensional centrally symmetric convex bodies which can be arbitrarily closely approximated by Minkowski sums of finitely many segments is a closed proper subset of $\mathcal{C}$. From this, our lemma follows. □

Verification of Schneider's Phenomenon: To check the positive part of Schneider's phemonemon first, consider the special case where D is a sum of l centrally symmetric segments with lengths η_i and directions u_i. Obviously,

$$h_D(u) = \frac{1}{2}\sum_{i=1}^{l} \eta_i |\langle u, u_i\rangle|.$$

Thus, by Lemma 2.2, Remark 2.1, and Corollary 2.1, we get

$$\begin{aligned}
n\, v(D)^{\frac{n-1}{n}} & \left(v(D)^{\frac{1}{n}} - v(C)^{\frac{1}{n}}\right) \\
&\geq n\,(v(D) - V_1(D,C)) \\
&= \int_{\partial(B)} h_D(u) G(D,du) - \int_{\partial(B)} h_D(u) G(C,du) \\
&= \frac{1}{2}\int_{\partial(B)} \sum_{i=1}^{l} \eta_i |\langle u, u_i\rangle| \Big(G(D,du) - G(C,du)\Big) \\
&= \frac{1}{2}\sum_{i=1}^{l} \eta_i \int_{\partial(B)} |\langle u, u_i\rangle| \Big(G(D,du) - G(C,du)\Big) \\
&= \sum_{i=1}^{l} \eta_i \Big(\bar{v}(P_u(D)) - \bar{v}(P_u(C))\Big).
\end{aligned}$$

Hence,

$$v(D) > v(C)$$

whenever $\bar{v}(P_u(D)) > \bar{v}(P_u(C))$ for every unit vector u.

In the general case, this last assumption clearly implies that for a suitable $\lambda > 1$ and all unit vectors u,

$$\bar{v}(P_u(D)) > \bar{v}(P_u(\lambda C)).$$

Thus, by modifying the previous argument a little through the use of the relation

$$V_1(D, \lambda C) = \lambda V_1(D,C) > V_1(D,C),$$

we see that the general case of the positive part of Schneider's phemonemon follows immediately by approximation.

Now let C_1 be as assumed in the negative part of Schneider's phenomenon. Lemma 6.9 implies that

$$h_{C_1}(u) = \int_{\partial(B)} |\langle u, w\rangle| f(w) d\lambda(w) \tag{6.46}$$

for some continuous function $f(w)$ on $\partial(B)$ satisfying $f(-w) = f(w)$.

If $f(w)$ were a non-negative function on $\partial(B)$, then the integral of (6.46) would be uniformly approximable by a sequence of finite sums

$$\sum |\langle u, u_i \rangle| f(u_i) \int_{A_i} d\lambda(w),$$

where each $u_i \in A_i \subset \partial(B)$. Since the supporting function $h_{C_1}(u)$ would then be arbitrarily approximable by linear forms

$$\sum \lambda_i |\langle u, u_i \rangle|$$

with suitable positive numbers λ_i and unit vectors u_i, one would then have that C_1 is arbitrarily closely approximable by Minkowski sums of finitely many segments, which is not the case by hypothesis.

Thus, $f(w)$ takes on negative values, and thus we may define a nonzero $k+1$ times differentiable function $G(u)$ on $\partial(B)$ satisfying $G(-w) = G(w)$ such that

$$G(u) = \begin{cases} \geq 0 & \text{if } f(u) < 0 \\ = 0 & \text{otherwise.} \end{cases} \tag{6.47}$$

Applying Lemma 6.9 once more, we get a continuous function $g(u)$ on $\partial(B)$ satisfying $g(-u) = g(u)$ such that

$$G(u) = \int_{\partial(B)} |\langle u, w \rangle| g(w) d\lambda(w). \tag{6.48}$$

Since ψ_{C_1} is positive on $\partial(B)$, by choosing $c > 0$ sufficiently small, we can guarantee that the function

$$g_1(u) = \psi_{C_1}(u) + cg(u) \tag{6.49}$$

is positive. Obviously,

$$\int_{\partial(B)} g_1(u) u \, d\lambda(u) = o.$$

Hence, by Lemma 3.1 there exists a centrally symmetric convex body C_2 such that

$$G(C_2, W) = \int_W g_1(w) d\lambda(w). \tag{6.50}$$

On the one hand, by Corollary 2.1, (6.50), and (6.49), we get

$$\begin{aligned} &\bar{v}(P_u(C_2)) - \bar{v}(P_u(C_1)) \\ &= \frac{1}{2} \int_{\partial(B)} |\langle u, w \rangle| G(C_2, dw) - \frac{1}{2} \int_{\partial(B)} |\langle u, w \rangle| G(C_1, dw) \\ &= \frac{1}{2} \int_{\partial(B)} |\langle u, w \rangle| \Big(g_1(w) - \psi_{C_1}(w) \Big) d\lambda(w) \\ &= \frac{c}{2} \int_{\partial(B)} |\langle u, w \rangle| g(w) d\lambda(w) \\ &= \frac{c}{2} G(u) \geq 0. \end{aligned} \tag{6.51}$$

On the other hand, by Lemma 2.2, Remark 2.1, (6.50), (6.46), (6.49), (6.48), and (6.47), we get

$$\begin{aligned}
& n\,v(C_1)^{\frac{n-1}{n}}\left(v(C_1)^{\frac{1}{n}} - v(C_2)^{\frac{1}{n}}\right) \\
& \geq n\,(v(C_1) - V_1(C_1, C_2)) \\
& = \int_{\partial(B)} h_{C_1}(w) G(C_1, dw) - \int_{\partial(B)} h_{C_1}(w) G(C_2, dw) \\
& = \int_{\partial(B)} h_{C_1}(w)\Big(\psi_{C_1}(w) - g_1(w)\Big) d\lambda(w) \\
& = -c \int_{\partial(B)} \Big\{\int_{\partial(B)} |\langle u, w\rangle| f(w) d\lambda(w)\Big\} g(u) d\lambda(u) \\
& = -c \int_{\partial(B)} \Big\{\int_{\partial(B)} |\langle u, w\rangle| g(u) d\lambda(u)\Big\} f(w) d\lambda(w) \\
& = -c \int_{\partial(B)} G(w) f(w) d\lambda(w) > 0,
\end{aligned}$$

which implies that

$$v(C_1) > v(C_2).$$

Since the $\geq$ of (6.51) can be changed into a $>$ by enlarging C_2 a little and the existence of C_1 is guaranteed by Lemma 6.10, the verification of Schneider's phenomenon is complete. □

§6. Some Historical Remarks

As was pointed out in the introduction, the history of the search for relationships between a convex body and its projections or sections is long and rich. Here, we restrict our attention to major developements connected with the Busemann-Petty problem.

The work of H. Busemann [1], which contains Corollary 6.1, should be regarded as the first remarkable developement since it is not only the forerunner of the Busemann-Petty problem but also a proper counterpart to the Larman-Rogers phenomenon.

The Busemann-Petty problem was first formulated in 1956. Four years later, H. Busemann [3] proved that the two hypotheses of convexity and symmetry are necessary. In other words, he found that the answer of the problem is "no" when either one of these hypotheses is dropped.

Meanwhile, the analogous problem for projections was raised by G.C. Shephard [1]. At the time, this problem was unsolved even for the convex centrally symmetric case, a characteristic it shared in common with the Busemann-Petty problem. However, although Shephard's problem and the Busemann-Petty problem appear to be closely related, the similarity is one of accidental appearance. A few years after being posed, C.M. Petty [1] achieved some partial results on Shephard's problem while, simultaneously, R. Schneider [1] completely solved it through the use of some deep results of Convex Geometry and Analysis, such as the Alexandrov-Fenchel inequality, the theory of spherical harmonics, and the solution of *Minkowski's problem* concerning area functions.

In 1975, D. Larman and C.A. Rogers [1] gave a negative solution to the Busemann-Petty problem for $n \geq 12$ by using some deep and complicated probabilistic arguments. Later, improvements of this result and simplifications of its proof were made by K. Ball, A.A. Giannopoulos, and J. Bourgain. More specifically, in 1988 K. Ball [1] showed that the Larman-Rogers phenomenon holds in R^n for $n \geq 10$ by studying the crossing sections of the cube; a few years later, J. Bourgain [1] and A.A. Giannopoulos [1] reduced the dimension here still further from 10 to 7.

In 1988, by developing a theory of dual mixed volumes and applying a technique of C.M. Petty and R. Schneider, E. Lutwak [1] was able to generalize Corollary 6.1 from ellipsoids to *intersection bodies*. Using Lutwak's theory, R.J. Gardner, M. Papadimitrakis, and G. Zhang recently proved a characterization theorem which R.J. Gardner [1] and M. Papadimitrakis [1] used to show that the Larman-Rogers phenomenon holds in R^5 and R^6. Then, G. Zhang [1] showed that the Larman-Rogers phenomenon holds in R^4, whereas R.J. Gardner [2] proved that it does not hold in R^3. Obviously, Gardner's final result makes the Larman-Rogers phenomenon even stranger.

7

Dvoretzky's Theorem

§1. Introduction

Definition 7.1. *Given a number ϵ strictly between 0 and 1, an n-dimensional convex set K is called an ϵ-sphere if there exists a positive number r such that*

$$rB \subseteq K \subseteq r(1+\epsilon)B.$$

In 1953, A. Grothendieck [1] made a conjecture in Functional Analysis which has as a consequence the following counterintuitive geometric assertion:

Given a positive integer n and a number ϵ strictly between 0 and 1, there exists an integer $M(n, \epsilon)$ such that for any m-dimensional centrally symmetric convex body C where $m \geq M(n, \epsilon)$, there exists an n-dimensional subspace R^n of R^m such that $C \cap R^n$ is an ϵ-sphere.

Although this statement seems rather improbable, it was proved in 1961 by A. Dvoretzky [1], and since then has become well known as *Dvoretzky's theorem.*

§2. Preliminaries

This section deals with a general result from Measure Theory and a theorem about sections of ellipsoids, both of which will play an important role in the proof of Dvoretzky's theorem.

Lemma 7.1. *Let $\{\mathfrak{R}, \rho\}$ be a compact metric space and let G be a group such that*

1. $\rho(gs, gt) = \rho(s,t)$ *for all* $g \in G$, $s, t \in \mathfrak{R}$
2. *for any distinct points* $s, t \in \mathfrak{R}$, *there exists a* $g \in G$ *such that* $gt = s$.

Then there exists a unique G-invariant probability measure μ on the Borel subsets of $\mathfrak{R}$.

Proof: Instead of proving the conclusion of the lemma as stated, we establish the following equivalent assertion:

Let $\mathcal{F}$ be the linear space of all real-valued continuous functions defined on $\mathfrak{R}$. Then there exists a unique measure μ such that

$$\mu(f) = \int_{\mathfrak{R}} f(t) d\mu(t) = \int_{\mathfrak{R}} f(gt) d\mu(t)$$

for all $f \in \mathcal{F}$ and $g \in G$, and $\mu(1) = 1$.

First, we show the existence of μ. For each $\epsilon > 0$, let N_ϵ be a *minimal ϵ-net* in $\mathfrak{R}$. In other words, let N_ϵ be a subset of points of $\mathfrak{R}$ such that

$$\bigcup_{s \in N_\epsilon} B(s, \epsilon) = \mathfrak{R},$$

with $n_\epsilon = \text{card}\{N_\epsilon\}$ being minimal among all sets with this property. (Here, $B(s,\epsilon)$ denotes $\{t \in \mathfrak{R} : \ \rho(t,s) \le \epsilon\}$, the usual closed ball centered at s of radius ϵ.) Then, for $f \in \mathcal{F}$ we may, and do, define

$$\mu_\epsilon(f) = n_\epsilon^{-1} \sum_{s \in N_\epsilon} f(s).$$

Since $\{\mu_\epsilon : \ \epsilon > 0\}$ is a family of uniformly bounded *linear functionals* on $\mathcal{F}$, it follows that for some sequence $\epsilon(i) \to 0$, we have

$$\mu_{\epsilon(i)}(f) \to \mu(f)$$

for all $f \in \mathcal{F}$ where μ is a positive linear functional on $\mathcal{F}$ satisfying $\mu(1) = 1$.

Next, we show that the measure μ is well defined by showing that if μ'_ϵ is defined by employing a different minimal ϵ-net N'_ϵ in $\mathfrak{R}$, then $\mu'_{\epsilon(i)}(f) \to \mu(f)$ for the same sequence $\epsilon(i)$.

We claim the following: *There is a one-to-one mapping $\Psi : N_\epsilon \mapsto N'_\epsilon$ such that $\rho(s, \Psi(s)) \le 2\epsilon$ for all $s \in N_\epsilon$.* To show this, we use a combinatorical result known as the *Marriage Lemma.* Let us say that $s \in N_\epsilon$ and $t \in N'_\epsilon$ are acquainted if $B(s,\epsilon) \cap B(t,\epsilon) \ne \emptyset$. Then the members of any subset A of N_ϵ are collectively acquainted with a subset A' of N'_ϵ of at least as many elements as A, for otherwise $A' \cup (N_\epsilon \setminus A)$ would be an ϵ-net with fewer than $\text{card}\{N_\epsilon\}$ elements. The marriage lemma says that in such a situation there is a one-to-one mapping $\Psi : N_\epsilon \mapsto N'_\epsilon$ such that s and $\Psi(s)$ are acquainted for all $s \in N_\epsilon$. This implies that $\rho(s, \Psi(s)) \leq 2\epsilon$ for all $s \in N_\epsilon$, and so

$$|\mu_\epsilon(f) - \mu'_\epsilon(f)| \leq n_\epsilon^{-1} \sum_{s \in N_\epsilon} |f(s) - f(\Psi(s))|$$
$$\leq \sup_{\rho(s,t) \leq 2\epsilon} \{|f(s) - f(t)|\}.$$

Clearly, then $\lim_{i \to +\infty} \mu'_{\epsilon(i)}(f)$ exists and is equal to $\mu(f)$, making μ well defined.

Second, we prove that μ is G-invariant. Letting $g \in G$, set $N'_\epsilon = \{gs : s \in N_\epsilon\}$ and note that N'_ϵ is a minimal ϵ-net. Thus,

$$\mu(f \circ g) = \lim_{i \to +\infty} \frac{1}{n_{\epsilon(i)}} \sum_{s \in N_{\epsilon(i)}} f(gs) = \lim_{i \to +\infty} \frac{1}{n_{\epsilon(i)}} \sum_{t \in N'_{\epsilon(i)}} f(t)$$
$$= \lim_{i \to +\infty} \mu'_{\epsilon(i)}(f) = \mu(f).$$

Third, we show the uniqueness of μ. Defining a *semimetric* ρ^* on G by $\rho^*(g,h) = \sup_{s \in \Re} \rho(gs, hs)$ and identifying elements whose distance apart is zero, we get a compact group H such that $\rho^*(gh_1, gh_2) = \rho^*(h_1, h_2)$ for $g \in G$ and $h_1, h_2 \in H$. Applying the existence assertions already proven, we see that there exists a G-invariant probability measure ν on H. Then, for all $f \in \mathcal{F}$, we have

$$\mu(f) = \nu(1)\mu(f) = \int_H \int_\Re f(hs) d\mu(s)\, d\nu(h)$$
$$= \int_\Re \int_H f(hs) d\nu(h)\, d\mu(s).$$

By the second hypothesis of the lemma and the G-invariance of ν, the inner integral on the right-hand side depends on f but not on s. Calling it $\nu'(f)$, we then have

$$\mu(f) = \nu'(f)\mu(1) = \nu'(f).$$

The same argument works for any other G-invariant probability measure μ' on $\Re$, so

$$\mu'(f) = \nu'(f) = \mu(f).$$

Hence, the proof of Lemma 7.1 is complete. □

We now state and prove an interesting result about the structure of ellipsoids.

Lemma 7.2. *Let E be an ellipsoid in R^{2l-1} centered at o. Then there is an l-dimensional subspace R^l such that $E \cap R^l$ is an l-dimensional sphere.*

Proof: Without loss of generality, we may assume that the ellipsoid under consideration is given by

$$\sum_{i=1}^{2l-1} r_i^2 \left(x^i\right)^2 \leq 1, \quad r_1 > r_2 > \cdots > r_{2l-1} > 0,$$

in a coordinate system with orthonormal basis $e_1, e_2, \ldots, e_{2l-1}$. Define $l-1$ positive numbers $\eta_1, \eta_2, \ldots, \eta_{l-1}$ by

$$\frac{r_i^2 \eta_i^2 + r_{2l-i}^2}{\eta_i^2 + 1} = r_l^2.$$

Then the l-dimensional subspace R^l of R^{2l-1} defined by the equations $x^i = \eta_i x^{2l-i}$, $1 \leq i \leq l-1$, has the desired property. Indeed, if we take

$$e_i' = \frac{\eta_i e_i + e_{2l-i}}{(\eta_i^2 + 1)^{\frac{1}{2}}}, \quad 1 \leq i \leq l-1,$$

and $e_l' = e_l$, then $\{e_1', e_2', \ldots e_l'\}$ is an orthonormal basis of R^l with respect to which the equation of $E \cap R^l$ is

$$r_l^2 \left(y^l\right)^2 + \sum_{i=1}^{l-1} \frac{r_i^2 \eta_i^2 + r_{2l-i}^2}{\eta_i^2 + 1} \left(y^i\right)^2 = \sum_{i=1}^{l} r_l^2 \left(y^i\right)^2 \leq 1.$$

This shows that $E \cap R^l$ is a sphere. □

§3. Technical Introduction

Given C a centrally symmetric convex body in R^l, we call

$$\|x\|_C = \begin{cases} 0, & x = o \\ \frac{\|x\|}{\|C(x)\|}, & x \neq o, \end{cases}$$

where $C(x)$ denotes the boundary point of C in the direction of x, the *Minkowski norm* on R^l determined by C. Then it is easy to see that C is an ϵ-sphere if and only if there exists a positive number r such that

$$r\|x\|_C \leq \|x\| \leq r(1+\epsilon)\|x\|_C$$

for all $x \in R^l$.

Because of Corollary 4.1, it is sufficient to prove Dvoretzky's theorem only for the regular centrally symmetric convex bodies. For convenience, we denote the family of all the regular l-dimensional centrally symmetric convex bodies by $\mathcal{C}^*$. For any fixed $C^* \in \mathcal{C}^*$, let

$$T = T_{C^*} : \partial(B^l) \mapsto \partial(B^l)$$

be the mapping such that $T(x)$ is the external unit normal to C^* at $C^*(x)$. Then we have

$$\left\langle \frac{x}{\|x\|_{C^*}}, T(x) \right\rangle = \sup_{y \in R^l} \left\{ \left\langle \frac{y}{\|y\|_{C^*}}, T(x) \right\rangle \right\}.$$

It is evident that $T(x) = x$ for all $x \in \partial(B^l)$ if and only if C^* is a sphere. It is also intuitively clear that if T is not far from the identity mapping, then C^* is an ϵ-sphere for some small ϵ.

Given R a subspace of R^l, let $\alpha_R(x)$ be the angle between x and $T_{C^* \cap R}(x)$ and set

$$\phi_R(x) = \langle x, T_{C^* \cap R}(x) \rangle.$$

Denote the *Stiefel manifold* of all pairs of orthonormal vectors in R by Ω_R. According to Lemma 7.1, there exists a unique rotation invariant probability measure σ_R on Ω_R. For $\{x, y\} \in \Omega_R$, set

$$\phi_y(x) = \phi_{H(x,y)}(x),$$

where $H(x, y)$ denotes the plane determined by o, x, and y. It is obvious that

$$\phi_R(x) = \inf\{\phi_y(x) : \{x, y\} \in \Omega_R\}. \tag{7.1}$$

For a fixed small positive number β, we define

$$A_R = \{\{x, y\} \in \Omega_R : \phi_y(x) > (1 - \beta^2)^{\frac{1}{2}}\}. \tag{7.2}$$

To fix notation, let R be an n-dimensional proper subspace of $V = R^k$ and let Γ be the *Grassmann manifold* of all n-dimensional subspaces of V. According to Lemma 7.1, there is a unique rotation invariant probability measure ϱ on Γ. By the uniqueness of σ_V, we have

$$\sigma_V(A_V) = \int_\Gamma \sigma_R(A_R) d\varrho(R) \tag{7.3}$$

and

$$\sigma_V(A_V) = \int_{\partial(B^k)} \lambda_x(A_x) d\lambda(x), \tag{7.4}$$

where

$$A_x = \{y \in \partial(B^k) : \{x, y\} \in A_V\}$$

and λ and λ_x are the normalized Lebesgue measures on $\partial(B^k)$ and $\{y \in \partial(B^k) : \langle x, y \rangle = 0\}$, respectively.

The proof of Dvoretzky's theorem that we will present is based on various estimates of $\sigma_V(A_V)$, $\sigma_R(A_R)$, and $\lambda_x(A_x)$.

§4. A Lemma of Dvoretzky and Rogers

Lemma 7.3 (A. Dvoretzky and C.A. Rogers [1]). *For any m-dimensional centrally symmetric convex body C where $m = 4k^2$, there exists an affine transformation L of R^m leaving the origin fixed and carrying C into a new convex body $L(C)$ such that*

1. $B \subseteq L(C)$
2. *there exists a k-dimensional subspace R^k such that $L(C) \cap R^k \subseteq 4W$.*

Proof: Let E be an ellipsoid that is maximal among those ellipsoids which can be inscribed in C. It is obvious that there exists an affine transformation L' such that $L'(o) = o$ and $L'(E) = B$. We now look more carefully at B and $L'(C)$.

Assertion 7.1. *There exist m points $p_1, p_2, \ldots, p_m$ and an orthogonal linear transformation L^* on R^m such that*

$$\{p_1, p_2, \ldots, p_m\} \subseteq \partial\left(L^*\left(L'(C)\right)\right) \cap \partial(B)$$

and such that for $j = 1, 2, \ldots, m$,

$$\begin{cases} p_j = (p_j^1, p_j^2, \ldots, p_j^j, 0, \ldots, 0), \\ \sum_{i=1}^{j-1} \left(p_j^i\right)^2 = 1 - \left(p_j^j\right)^2 \le \frac{j-1}{m}. \end{cases} \tag{7.5}$$

For $j = 1$, the existence of p_1 and the corresponding orthogonal linear transformation L_1^* is clear. Assuming that it is true for $j = l - 1 < m$, we now prove it for $j = l$. It is easy to see that for $\epsilon > 0$, the ellipsoid

$$(1+\epsilon)^{m-l+1} \sum_{i=1}^{l-1} (x^i)^2 + (1+\epsilon+\epsilon^2)^{-l+1} \sum_{i=l}^{m} (x^i)^2 \le 1 \tag{7.6}$$

has volume

$$\left(\frac{1+\epsilon+\epsilon^2}{1+\epsilon}\right)^{\frac{(m-l+1)(l-1)}{2}} \omega_m.$$

Therefore, there is a point $p(\epsilon) = (p^1, p^2, \ldots, p^m)$ on the boundary of $L'(C)$ and in the ellipsoid (7.6). Since $p(\epsilon)$ does not belong to the interior of the unit sphere, we have

$$\sum_{i=1}^{m} \left(p^i\right)^2 \ge 1,$$

and so

$$\left((1+\epsilon)^{m-l+1} - 1\right) \sum_{i=1}^{l-1} \left(p^i\right)^2 + \left((1+\epsilon+\epsilon^2)^{-l+1} - 1\right) \sum_{i=l}^{m} \left(p^i\right)^2 \le 0. \tag{7.7}$$

If $\epsilon \to 0$ through a suitable sequence of positive numbers, the corresponding sequence $p(\epsilon)$ will converge to a point p_l. It is clear from (7.6) that p_l lies on the boundaries of both $L'(C)$ and B, whereas from (7.7) we see that

$$(m-l+1)\sum_{i=1}^{l-1}\left(p_l^i\right)^2+(-l+1)\sum_{i=l}^{m}\left(p_l^i\right)^2 \leq 0. \tag{7.8}$$

By basic Linear Algebra, there exists an orthogonal transformation

$$L_l^*: \begin{array}{l} e_i' = e_i, \quad i=1,\ \ldots,\ l-1, \\ e_l' = \dfrac{1}{\sum_{i=l}^{m}\left(p_l^i\right)^2}\displaystyle\sum_{i=l}^{m} p_l^i e_i, \\ \vdots \end{array}$$

Hence, by (7.8) and the fact that $p_l \in \partial(B)$, we see that $p_1, \ldots, p_l$ and $\prod_{i=1}^{l} L_i^*$ satisfy Assertion 7.1 for $j=l$.

Our induction complete, Assertion 7.1 is proven by taking $L^* = \prod_{i=1}^{m} L_i^*$.

We now proceed to show that $L = L^*L'$ and $R^k = \{\sum_{i=1}^{k}\alpha_i p_i : -\infty < \alpha_i < +\infty\}$ satisfy the conclusion of our lemma.

Clearly, we have $B \subseteq L(C)$. By Assertion 7.1, we notice that the only supporting hyperplane of $L(C)$ at p_j, $1 \leq j \leq m$, is

$$H_j: \quad \sum_{i=1}^{j} p_j^i x^i = 1.$$

Thus, $L(C) \cap R^k$ is contained in the parallelepiped P defined by

$$\left|\sum_{i=1}^{j} p_j^i x^i\right| \leq 1, \quad j=1,\ 2, \ldots, k. \tag{7.9}$$

To complete our proof it suffices to show that $P \subseteq 4W$, i.e., that

$$(x^i)^2 \leq 4 \quad \text{whenever } (x^1, x^2, \ldots, x^k) \in P. \tag{7.10}$$

For $i=1$, (7.10) is clear. Assuming that (7.10) holds for $i = l-1 < k$, we now prove it for $i=l$. Since

$$\left(1+\frac{2(l-1)}{\sqrt{m}}\right)^2 < 4\left(1-\frac{l-1}{m}\right),$$

by (7.9) and (7.5) we see that

$$(x^l)^2 \leq \left(p_l^l\right)^{-2}\left(1+\sum_{j=1}^{l-1}\left|p_l^j\right|\cdot\left|x^j\right|\right)^2$$

$$\leq \left(p_l^l\right)^{-2}\Big\{1+\Big(\sum_{j=1}^{l-1}\left(p_l^j\right)^2\Big)^{\frac{1}{2}}\Big(\sum_{j=1}^{l-1}(x^j)^2\Big)^{\frac{1}{2}}\Big\}^2$$

$$\leq \left(1-\frac{l-1}{m}\right)^{-1}\Big\{1+\left(\frac{l-1}{m}\right)^{\frac{1}{2}}\sqrt{4(l-1)}\Big\}^2 \leq 4.$$

Thus, (7.10) is proven and the proof of Lemma 7.3 is complete. □

§5. An Estimate for $\sigma_V(A_V)$

On $\partial(B^k)$ we shall be employing the geodesic distance ρ. Let $\mu(k,\beta)$ be the λ measure of the β-neighborhood of a point in $\partial(B^k)$ and $\nu(k,\beta)$ be the λ measure of the β-neighborhood of an *equator* in $\partial(B^k)$.

Taking

$$I_k = \int_0^{\frac{\pi}{2}} \cos^k \theta d\theta = \int_0^{\frac{\pi}{2}} \sin^k \theta d\theta,$$

it is well known that

$$\sqrt{\frac{\pi k}{2(k+1)}} \leq I_k\sqrt{k} \leq \sqrt{\frac{\pi}{2}}.$$

Hence, we have

$$k^{\frac{1}{2}}\left(\frac{\beta}{4}\right)^{k-1} \leq \mu(k,\beta) = \frac{\int_0^\beta \sin^{k-2}\theta d\theta}{2I_{k-2}} \leq (k-1)^{\frac{1}{2}}\sin^{k-1}\beta. \tag{7.11}$$

Since

$$\cos^k \frac{\gamma}{\sqrt{k}} \leq e^{-\frac{\gamma^2}{2}}$$

for $k \geq 1$ and $0 \leq \gamma \leq \frac{\pi}{2}\sqrt{k}$, we also have

$$\begin{aligned}
\nu(k+2,\beta) &= 1 - \frac{1}{I_k}\int_\beta^{\frac{\pi}{2}} \cos^k\theta d\theta \\
&= 1 - \frac{1}{I_k\sqrt{k}}\int_{\beta\sqrt{k}}^{\frac{\pi}{2}\sqrt{k}} \cos^k\frac{\gamma}{\sqrt{k}} d\gamma \\
&\geq 1 - \sqrt{\frac{2(k+1)}{k\pi}}\int_{\beta\sqrt{k}}^{\frac{\pi}{2}\sqrt{k}} e^{-\frac{\gamma^2}{2}} d\gamma \\
&\geq 1 - \sqrt{\frac{4}{\pi}}e^{-\frac{\beta^2 k}{2}}\int_0^{+\infty} e^{-\frac{\gamma^2}{2}} d\gamma \\
&= 1 - \sqrt{2}e^{-\frac{\beta^2 k}{2}} \geq 1 - \beta^{-1}k^{-\frac{1}{2}}.
\end{aligned} \tag{7.12}$$

We now state and prove a fundamental result about the global integral of the reciprocal of the Minkowski norm determined by W^k.

Lemma 7.4 (T. Figiel [1]).

$$\int_{\partial(B^k)} \|x\|_{W^k}^{-1} d\lambda(x) = O\left(\frac{k}{2\log k}\right)^{\frac{1}{2}}.$$

Proof: In R^k, we define

$$f_k(x) = \pi^{-\frac{k}{2}} \|x\|_{W^k}^{-1} e^{-\|x\|^2}$$

and

$$J_k = \int_{R^k} f_k(x) dl(x),$$

where $l(x)$ denotes Lebesgue measure on R^k. Passing to spherical coordinates and using the well-known result that $k\omega_k = 2\pi^{\frac{k}{2}}\Gamma^{-1}(\frac{k}{2})$, we get

$$\begin{aligned} J_k &= \int_0^{+\infty} dt \int_{\partial(tB^k)} f_k(x) ds_t(x) \\ &= \pi^{-\frac{k}{2}} k\omega_k \int_{\partial(B^k)} \|x\|_{W^k}^{-1} d\lambda(x) \int_0^{+\infty} t^{k-2} e^{-t^2} dt \\ &= \Gamma\left(\frac{k-1}{2}\right) \Gamma^{-1}\left(\frac{k}{2}\right) \int_{\partial(B^k)} \|x\|_{W^k}^{-1} d\lambda(x), \end{aligned}$$

where $s_t(x)$ denotes area measure on $\partial(tB^k)$.

Since $\Gamma\left(\frac{k}{2}\right)\Gamma^{-1}\left(\frac{k-1}{2}\right) \sim \sqrt{\frac{k}{2}}$, to prove the lemma it suffices to show that $J_k = O\left(\frac{1}{\sqrt{\log k}}\right)$. For

$$\Psi(t) = \int_{tW^k} f_k(x) dl(x)$$

and

$$\psi(t) = \pi^{-\frac{k}{2}} \int_{tW^k} e^{-\|x\|^2} dl(x) = \left(\frac{2}{\sqrt{\pi}} \int_0^t e^{-\gamma^2} d\gamma\right)^k,$$

one has that $\Psi'(t) = t^{-1}\psi'(t)$ for $t > 0$. Integrating by parts, we obtain

$$\begin{aligned} J_k &= \int_0^{+\infty} d\Psi(t) = \int_0^{+\infty} t^{-1} d\psi(t) = \int_0^{+\infty} t^{-2}\psi(t) dt \\ &= \int_0^{+\infty} t^{-2} \left(\frac{2}{\sqrt{\pi}} \int_0^t e^{-\gamma^2} d\gamma\right)^k dt. \end{aligned}$$

Applying the simple inequality

$$\int_0^t e^{-\gamma^2} d\gamma \leq \min\left\{\frac{\sqrt{\pi}}{2}, \frac{2t}{t+1}\right\},$$

valid for $t > 0$, we see that for any positive ξ,

$$\begin{aligned} J_k &= \int_0^{+\infty} t^{-2}\left(\frac{2}{\sqrt{\pi}}\int_0^t e^{-\gamma^2} d\gamma\right)^k dt = \int_0^{\xi} + \int_{\xi}^{+\infty} \\ &\leq \frac{16}{\pi}\int_0^{\xi} (t+1)^{-2}\left(\frac{2}{\sqrt{\pi}}\int_0^t e^{-\gamma^2} d\gamma\right)^{k-2} dt + \int_{\xi}^{+\infty} t^{-2} dt \\ &= \frac{8}{(k-1)\sqrt{\pi}}\int_0^{\xi} (t+1)^{-2} e^{t^2} \frac{d}{dt}\left\{\left(\frac{2}{\sqrt{\pi}}\int_0^t e^{-\gamma^2} d\gamma\right)^{k-1}\right\} dt + \xi^{-1} \\ &\leq \frac{5}{k-1}\sup_{0\leq t\leq \xi}\{(t+1)^{-2}e^{t^2}\} + \xi^{-1}. \end{aligned}$$

Since

$$\sup_{0\leq t\leq \xi}\{(t+1)^{-2}e^{t^2}\} = \max\{1, (\xi+1)^{-2}e^{\xi^2}\},$$

by taking $\xi = \sqrt{\log k}$, we obtain

$$J_k \leq \frac{1}{\sqrt{\log k}} + \frac{5k}{(k-1)\log k} = O\left(\frac{1}{\sqrt{\log k}}\right),$$

which completes the proof of Lemma 7.4. □

Lemma 7.5. *Suppose that $B \subseteq C^* \subset 4W \subset R^k$. Then*

$$\sigma_V(A_V) > 1 - 8\beta^{-1}k^{-\frac{1}{2}} \int_{\partial(B)} \|x\|_W^{-1} d\lambda(x). \tag{7.13}$$

Proof: Let $x \in R \subset V$ be a unit vector and let Υ be the orthogonal projection from V to R. Then

$$\Upsilon\left(T(x)\right) = \|\Upsilon\left(T(x)\right)\| \cdot T_{C^*\cap R}(x),$$

and consequently,

$$\phi_V(x) = \|\Upsilon\left(T(x)\right)\| \cdot \phi_R(x).$$

If $\{x, y\} \in \Omega_V$ and $R = H(x, y)$, then

$$\Upsilon\left(T(x)\right) = \langle x, T(x)\rangle x + \langle y, T(x)\rangle y = \phi_V(x)x + \langle y, T(x)\rangle y,$$

and so, finally,

$$\phi_y(x) = \phi_V(x)\left(\phi_V^2(x) + \langle y, T(x)\rangle^2\right)^{-\frac{1}{2}}. \tag{7.14}$$

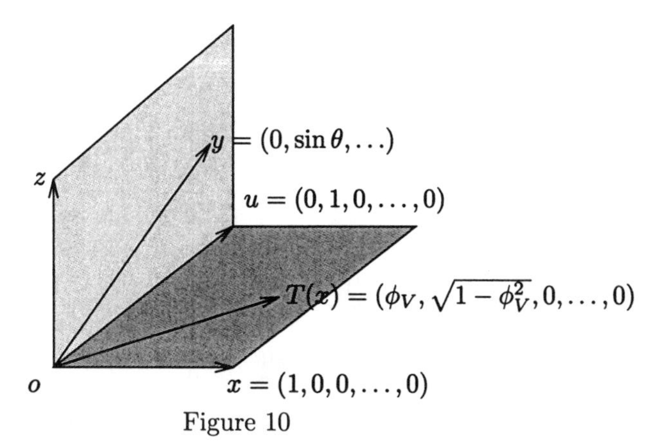

Figure 10

Letting θ be the angle between y and the $(k-2)$-dimensional hyperplane $\{z \in V : \langle z, x\rangle = 0, \langle z, T(x)\rangle = 0\}$, we have that

$$\phi_y(x) = \phi_V(x)\Big(\phi_V^2(x) + (1 - \phi_V^2(x))\sin^2\theta\Big)^{-\frac{1}{2}}.$$

Thus, $\phi_y(x) > (1-\beta^2)^{\frac{1}{2}}$ is satisfied, provided that

$$\Big(1 + \sin^2\theta\tan^2\alpha\Big)^{-1} > 1 - \beta^2, \tag{7.15}$$

where $\alpha = \arccos\phi_V(x)$. Clearly, (7.15) is satisfied when

$$\theta < |\beta\cot\alpha|.$$

From this and (7.12), it follows that for $k \geq 6$,

$$\lambda_x(A_x) \geq \nu(k-1, |\beta\cot\alpha|) \geq 1 - 2\beta^{-1}k^{-\frac{1}{2}}|\tan\alpha|. \tag{7.16}$$

By hypothesis, we have

$$\sup_{y\in\partial(C^*)}\{\langle y, T(x)\rangle\} \geq \|T(x)\| = 1,$$

and hence

$$\begin{aligned}|\cot\alpha| > \cos\alpha = \phi_V(x) &= \langle x, T(x)\rangle \\ &\geq \frac{\langle x, T(x)\rangle}{\sup_{y\in\partial(C^*)}\{\langle y, T(x)\rangle\}} = \|x\|_{C^*} \geq \frac{1}{4}\|x\|_W.\end{aligned} \tag{7.17}$$

So by (7.16) and (7.17), we obtain

$$\lambda_x(A_x) \geq 1 - 8\beta^{-1}k^{-\frac{1}{2}}\|x\|_W^{-1}$$

and

$$\sigma_V(A_V) = \int_{\partial(B)} \lambda_x(A_x)d\lambda(x) \geq 1 - 8\beta^{-1}k^{-\frac{1}{2}}\int_{\partial(B)}\|x\|_W^{-1}d\lambda(x).$$

Lemma 7.5 is proven. □

§6. β-nets and ϵ-spheres

As usual, a set $\{z_1, z_2, \ldots, z_g\}$ is called a *β-net* in a compact metric space $\{\Re, \rho\}$ if

$$\Re = \bigcup_{i=1}^{g} \{x \in \Re : \ \rho(x, z_i) \leq \beta\}.$$

In this section, we will restrict ourselves to the metric space $\Re = \partial(B^l)$ equipped with its geodesic distance ρ. For convenience, we use R and B as abbreviations for R^l and B^l, respectively.

Lemma 7.6. *Let β and η be small positive numbers. If $\Im$ is a β-net in $\partial(B)$ such that*

$$(1+\eta) \inf_{z \in \Im} \{\|z\|_{C^*}\} \geq \sup_{z \in \Im} \{\|z\|_{C^*}\} = r,$$

then C^* is a $(2\eta + 5\beta)$-sphere.

Proof: Without loss of generality, we may take $r = 1$. On the one hand, by hypothesis, we see that for every $x \in \partial(B)$,

$$\begin{aligned}\frac{\langle x, T(x)\rangle}{\|x\|_{C^*}} &= \sup_{y \in \partial(C^*)} \{\langle y, T(x)\rangle\} \geq \sup_{z \in \Im} \left\{ \frac{\langle z, T(x)\rangle}{\|z\|_{C^*}} \right\} \\ &\geq \sup_{z \in \Im} \{\langle z, T(x)\rangle\} \geq \cos\beta \geq (1-\beta^2)^{\frac{1}{2}}.\end{aligned}$$

Therefore, we have

$$\|x\|_{C^*} \leq (1-\beta^2)^{-\frac{1}{2}}. \tag{7.18}$$

On the other hand, letting $z \in \Im$ be a point such that $\rho(x, z) \leq \beta$, we have

$$\begin{aligned}\|x\|_{C^*} &\geq \frac{\langle x, T(z)\rangle}{\sup_{y \in \partial(C^*)}\{\langle y, T(z)\rangle\}} = \|z\|_{C^*} + \frac{\langle x - z, T(z)\rangle}{\sup_{y \in \partial(C^*)}\{\langle y, T(z)\rangle\}} \\ &\geq (1+\eta)^{-1} - \|x - z\| \cdot (1-\beta^2)^{-\frac{1}{2}} \\ &\geq 1 - \eta - 2\beta(1-\beta^2)^{-\frac{1}{2}} \geq 1 - \eta - 3\beta.\end{aligned} \tag{7.19}$$

Clearly, (7.18) and (7.19) imply that C^* is an ϵ-sphere with

$$\begin{aligned}\epsilon &= (1-\beta^2)^{-\frac{1}{2}}(1-\eta-3\beta)^{-1} - 1 \leq (1+\beta)(1+\eta+3\beta) - 1 \\ &\leq 2\eta + 5\beta.\end{aligned}$$

Lemma 7.6 is proven. □

Lemma 7.7. *Take $\beta = 10^{-3}\epsilon$. If the set*

$$\Im = \{z \in \partial(B) : \ \phi_R(z) > (1 - 6^2\beta^2)^{\frac{1}{2}}\}$$

is a β-net in $\partial(B)$, then C^* is an ϵ-sphere.

Proof: Let $z' \in \Im$ be a point such that

$$\|z'\|_{C^*} = \sup_{z \in \Im}\{\|z\|_{C^*}\}.$$

Clearly, for any point $z \in \Im$, there exists a sequence $z_0 = \pm z', z_1, \ldots, z_h = z \in \Im$ such that

$$h \leq \frac{\pi}{2}\beta^{-1}$$

and

$$\rho(z_i, z_{i+1}) \leq 3\beta, \quad i = 0,\ 1, \ldots, h-1.$$

Since

$$\rho(z_i, T(z_i)) \leq 2(1 - \phi_R^2(z_i))^{\frac{1}{2}} < 12\beta,$$

we have for $i = 0,\ 1, \ldots, h-1$,

$$\rho(z_{i+1}, T(z_i)) \leq \rho(z_{i+1}, z_i) + \rho(z_i, T(z_i)) < 15\beta.$$

Hence,

$$\begin{aligned}\|z_{i+1}\|_{C^*} &\geq \frac{\langle z_{i+1}, T(z_i)\rangle}{\sup_{y \in \partial(C^*)}\{\langle y, T(z_i)\rangle\}} \\ &\geq (1 - 15^2\beta^2)^{\frac{1}{2}}\|z_i\|_{C^*}\phi_R^{-1}(z_i) \\ &\geq (1 - 15^2\beta^2)^{\frac{1}{2}}\|z_i\|_{C^*}.\end{aligned}$$

Therefore, applying the inequality $e^t \leq 1 + 2t$ for $0 \leq t \leq 1$, we obtain

$$\begin{aligned}\|z\|_{C^*} &\geq (1 - 15^2\beta^2)^{\frac{h}{2}}\|z'\|_{C^*} \geq (1 - 15^2\beta^2)^{\beta^{-1}}\|z'\|_{C^*} \\ &\geq e^{-225\beta}\|z'\|_{C^*} \geq (1 + 450\beta)^{-1}\|z'\|_{C^*}.\end{aligned}$$

Finally, applying Lemma 7.6 with $\eta = 450\beta$, we see that C^* is an ϵ-sphere. □

Lemma 7.8. *Let $\beta = 10^{-3}\epsilon$. If*

$$\sigma_R(A_R) > 1 - \mu(l, \beta)\mu(l-1, \beta),$$

then C^* is an ϵ-sphere.

Proof: Taking $\Im = \{z \in \partial(B) : \ \lambda_z(A_z) > 1 - \mu(l-1, \beta)\}$, by (7.4) and the hypothesis we see that

$$\lambda(\Im) > 1 - \mu(l, \beta),$$

and, hence, $\Im$ is a β-net in $\partial(B)$.

To apply Lemma 7.7, we need to show that

$$\phi_R(z) > (1 - 6^2\beta^2)^{\frac{1}{2}} \tag{7.20}$$

for every $z \in \Im$.

Letting $z \in \Im$, it is easy to see that A_z is a β-net in $\{x \in \partial(B) : \langle z, x\rangle = 0\}$. Let x be an arbitrary point of $\{x \in \partial(B) : \langle x, z\rangle = 0\}$ and let $y \in A_z$ be a point such that $\rho(y, x) < \beta$. Then $\|y - x\| < \beta$. Assuming that $\beta < \frac{1}{4}$, it then follows from (7.14) and (7.2) that

$$\langle y, T(z)\rangle = \left(\frac{1}{\phi_y^2(z)} - 1\right)^{\frac{1}{2}} \phi_R(z) \le \left(\frac{1}{1-\beta^2} - 1\right)^{\frac{1}{2}} < 2\beta,$$

$$\begin{aligned}\phi_R(z) &= \max_{w \in A_z} \{\phi_w(z)(\phi_R^2(z) + \langle w, T(z)\rangle^2)^{\frac{1}{2}}\}\\ &\ge \sqrt{1-\beta^2}\sqrt{1-\beta^2} > \tfrac{1}{2},\end{aligned}$$

and

$$\begin{aligned}\langle x, T(z)\rangle = \langle x - y, T(z)\rangle + \langle y, T(z)\rangle &\le \|x - y\| + 2\beta < 3\beta\\ &< 6\beta\phi_R(z).\end{aligned}$$

Hence, by (7.14) we get

$$\begin{aligned}\phi_x(z) &= \frac{\phi_R(z)}{(\phi_R^2(z) + \langle x, T(z)\rangle^2)^{\frac{1}{2}}} > \frac{\phi_R(z)}{(1 + 6^2\beta^2)^{\frac{1}{2}}\phi_R(z)}\\ &> (1 - 6^2\beta^2)^{\frac{1}{2}}.\end{aligned}$$

Thus, (7.20) follows from (7.1).

Lemma 7.8 now follows immediately from Lemma 7.7. □

§7. A Proof of Dvoretzky's Theorem

Theorem 7.1 (A. Dvoretzky [1]). *Given a positive integer n and a number ϵ strictly between 0 and 1, there exists an integer $M(n, \epsilon)$ such that for any m-dimensional centrally symmetric convex body C where $m \ge M(n, \epsilon)$, there exists an n-dimensional subspace R^n of R^m such that $C \cap R^n$ is an ϵ-sphere.*

Proof: By Corollary 4.1, it suffices to prove Dvoretzky's theorem only for the regular centrally symmetric convex bodies $C^* \in \mathcal{C}^*$.

First, by applying Lemma 7.3 to C^*, we get a linear transformation L and a proper subspace R^k of R^m such that

$$B^k \subseteq L(C^*) \cap R^k \subset 4W^k.$$

Then, applying Lemma 7.4 and Lemma 7.5 to $V = R^k$ where $k > 2n - 1$, we get

$$\sigma_V(A_V) > 1 - \frac{2 \log\log k}{\beta \log k}.$$

Keeping (7.3) in mind, we obtain

$$\int_\Gamma \sigma_R(A_R)d\varrho(R) > 1 - \frac{2\log\log k}{\beta\log k}.$$

Thus, according to Lemma 7.8, when $\beta = 10^{-3}\epsilon$ and

$$1 - \frac{2\log\log k}{\beta\log k} > 1 - \mu(2n-1,\beta)\cdot\mu(2n-2,\beta);$$

i.e., when $\beta = 10^{-3}\epsilon$ and

$$\frac{\log\log k}{\log k} < \frac{\beta}{2}\mu(2n-1,\beta)\cdot\mu(2n-2,\beta),$$

there is a subspace R^{2n-1} of R^m such that $L(C^*)\cap R^{2n-1}$ is an ϵ-sphere. In other words, $L^{-1}(R^{2n-1})\cap C^*$ is a $(2n-1)$-dimensional ϵ-ellipsoid. Finally, Lemma 7.2 implies the existence of an n-dimensional ϵ-spherical section of C^*.

Theorem 7.1 is proven. □

§8. An Upper Bound for $M(n,\epsilon)$

It follows from (7.16), (7.17), and (7.12) that

$$\lambda_x(A_x) \geq 1 - 2e^{-\frac{\beta^2 k}{16}\|x\|^2_{W^k}}.$$

Therefore, by (7.4)

$$\sigma_V(A_V) \geq 1 - 2\int_{\partial(B^k)} e^{-\frac{\beta^2 k}{16}\|x\|^2_{W^k}} d\lambda(x). \tag{7.21}$$

Thus, in obtaining a precise upper bound for $M(n,\epsilon)$, a key role is played by the estimation of the integral

$$I(k) = \int_{\partial(B^k)} e^{-\frac{\beta^2 k}{16}\|x\|^2_{W^k}} d\lambda(x).$$

This key estimation requires, in turn, the following result.

Lemma 7.9. *Let*

$$J(k) = \int_{\partial(B^k)} \|x\|^{-2g}_{W^k} d\lambda(x).$$

Then, for every number $g \leq k^{\frac{1}{8}}$,

$$J(k) \leq \left(\frac{k}{\log k}\right)^g (1+o(1))$$

Proof: We consider the function on R^k defined by

$$f(x) = \pi^{-\frac{k}{2}} e^{-\|x\|^2} \|x\|_{W^k}^{-2g}$$

and set

$$L(k) = \int_{R^k} f(x) dl(x).$$

Clearly, we have

$$\begin{aligned} L(k) &= \int_0^{+\infty} dt \int_{\partial(tB^k)} \pi^{-\frac{k}{2}} e^{-t^2} \|x\|_{W^k}^{-2g} ds_t(x) \\ &= \int_0^{+\infty} \pi^{-\frac{k}{2}} t^{-2g+k-1} e^{-t^2} dt \int_{\partial(B^k)} \|x\|_{W^k}^{-2g} ds_1(x), \end{aligned}$$

where $s_t(x)$ denotes area measure on $\partial(tB^k)$ and $ds_1(x) = k\omega_k d\lambda(x)$. Hence,

$$L(k) = \frac{J(k)}{\Gamma\left(\frac{k}{2}\right)} \int_0^{+\infty} 2t^{k-2g-1} e^{-t^2} dt = \frac{\Gamma\left(\frac{k}{2} - g\right)}{\Gamma\left(\frac{k}{2}\right)} J(k),$$

and therefore

$$J(k) = \frac{\Gamma\left(\frac{k}{2}\right)}{\Gamma\left(\frac{k}{2} - g\right)} L(k) \leq \left(\frac{k}{2}\right)^g \left(1 + o(1)\right) L(k). \tag{7.22}$$

In estimating $L(k)$, it is convenient to set

$$\Psi(t) = \int_{2tW^k} f(x) dl(x)$$

and

$$\psi(t) = \pi^{-\frac{k}{2}} \int_{2tW^k} e^{-\|x\|^2} dl(x) = \left(\frac{2}{\sqrt{\pi}} \int_0^t e^{-\gamma^2} d\gamma \right)^k.$$

Noting that $\Psi'(t) = t^{-2g} \psi'(t)$ and integrating by parts, we obtain

$$\begin{aligned} L(k) &= \int_0^{+\infty} \Psi'(t) dt = 2g \int_0^{+\infty} t^{-2g-1} \psi(t) dt \\ &= 2g \int_0^{+\infty} t^{-2g-1} \left(\frac{2}{\sqrt{\pi}} \int_0^t e^{-\gamma^2} d\gamma \right)^k dt. \end{aligned}$$

Taking $\xi = \left(\frac{\log k}{2}\right)^{\frac{1}{2}}$, we split up the last integral in the following way:

$$2g \int_0^{+\infty} t^{-2g-1} \left(\frac{2}{\sqrt{\pi}} \int_0^t e^{-\gamma^2} d\gamma \right)^k dt = 2g \left(\int_0^2 + \int_2^\xi + \int_\xi^{+\infty} \right).$$

We shall show that the first two integrals are of a smaller order of magnitude than the last one, which is estimated by

$$2g\int_{\xi}^{+\infty} \le 2g\int_{\xi}^{+\infty} t^{-2g-1}dt \le \xi^{-2g} = \left(\frac{\log k}{2}\right)^{-g}.$$

To estimate the first integral, set $\alpha = \frac{2}{\sqrt{\pi}}\int_0^2 e^{-\gamma^2}d\gamma$. For every t between 0 and 2, we have

$$\frac{2}{\sqrt{\pi}}\int_0^t e^{-\gamma^2}d\gamma \le \min\left\{\frac{2t}{\sqrt{\pi}}, \alpha\right\}.$$

Then, taking η to be a number such that $\alpha^\eta = \frac{\sqrt{\pi}}{2}$, note that we have

$$\begin{aligned}
2g\int_0^2 &= 2g\int_0^2 t^{-2g-1}\left(\frac{2}{\sqrt{\pi}}\int_0^t e^{-\gamma^2}d\gamma\right)^{2g+1}\left(\frac{2}{\sqrt{\pi}}\int_0^t e^{-\gamma^2}d\gamma\right)^{k-2g-1} dt \\
&\le 4g\left(\frac{2}{\sqrt{\pi}}\right)^{2g+1}\alpha^{k-2g-1} = O\left(\alpha^{k-3(\eta+1)g}\right) \\
&= o\left((\log k)^{-g}\right).
\end{aligned}$$

To estimate the second integral, observe that

$$\begin{aligned}
2g\int_2^{\xi} &\le 2g\int_2^{+\infty} t^{-2g-1}dt \cdot \left(\frac{2}{\sqrt{\pi}}\int_0^{\xi} e^{-\gamma^2}d\gamma\right)^k \\
&= 2^{-2g}\left(1 - \frac{2}{\sqrt{\pi}}\int_{\xi}^{+\infty} e^{-\gamma^2}d\gamma\right)^k \\
&\le 2^{-2g}\left(1 - \frac{2}{\sqrt{\pi}}\int_{\left(\frac{\log k}{2}\right)^{\frac{1}{2}}}^{\left(\frac{3\log k}{4}\right)^{\frac{1}{2}}} e^{-\gamma^2}d\gamma\right)^k \\
&\le 2^{-2g}\left\{1 - \frac{2}{\sqrt{\pi}}\left[\left(\frac{3}{4}\right)^{\frac{1}{2}} - \left(\frac{1}{2}\right)^{\frac{1}{2}}\right](\log k)^{\frac{1}{2}} e^{-\frac{3\log k}{4}}\right\}^k \\
&\le 2^{-2g}\left(1 - \frac{(\log k)^{\frac{1}{2}}k^{\frac{1}{4}}}{10k}\right)^k \le 2^{-2g}2^{-\frac{(\log k)^{\frac{1}{2}}k^{\frac{1}{4}}}{10}} \\
&= o\left(\left(\frac{\log k}{2}\right)^{-g}\right).
\end{aligned}$$

Thus, by (7.22) and the estimates on $2g\int_0^2$, $2g\int_2^{\xi}$, and $2g\int_{\xi}^{+\infty}$, we get

$$\begin{aligned}
J(k) &\le \left(\frac{k}{2}\right)^g\left(1+o(1)\right)\left\{o\left((\log k)^{-g}\right) + o\left(\left(\frac{\log k}{2}\right)^{-g}\right) + \left(\frac{\log k}{2}\right)^{-g}\right\} \\
&= \left(\frac{k}{\log k}\right)^g(1+o(1)).
\end{aligned}$$

Lemma 7.9 is proven. □

We now proceed to apply Lemma 7.9 to estimate $I(k)$.

Since the function $x^{2g}e^{-x^2}$ attains its maximum at $x = g^{\frac{1}{2}}$, we have

$$e^{-x^2} \leq g^g e^{-g} x^{-2g}$$

for all $g > 0$ and $x > 0$. Thus, we see that for every $g \leq k^{\frac{1}{8}}$,

$$\begin{aligned} I(k) &\leq g^g \left(\frac{16}{e}\right)^g \beta^{-2g} k^{-g} J(k) \\ &\leq 2g^g \left(\frac{16}{e}\right)^g \beta^{-2g} (\log k)^{-g} \Big(1 + o(1)\Big). \end{aligned}$$

Without loss of generality, in the argument that follows, we may omit the $o(1)$ term of the last equation. It is then obvious that the right-hand side of this equation attains its minimum at

$$g = \frac{1}{16}\beta^2 \log k,$$

and hence

$$I(k) \leq e^{-\frac{1}{16}\beta^2 \log k} = k^{-\frac{\beta^2}{16}}.$$

Thus, by (7.21), Lemma 7.8, Lemma 7.2, and (7.11), if

$$\begin{aligned} 1 - k^{-\frac{\beta^2}{16}} &\geq 1 - \mu(2n-1, 10^{-3}\epsilon) \cdot \mu(2n-2, 10^{-3}\epsilon) \\ &\geq 1 - 2n \left(\frac{\epsilon}{4 \times 10^3}\right)^{4n-4}, \end{aligned}$$

$$k \geq \left\{ 2n \left(\frac{\epsilon}{4 \times 10^3}\right)^{4n-4} \right\}^{-\frac{1.6 \times 10^7}{\epsilon^2}},$$

and (keeping Lemma 7.3 in mind)

$$\begin{aligned} m \geq M(n, \epsilon) &= 4k^2 \\ &\geq 4 \times (2n)^{-\frac{3.2 \times 10^7}{\epsilon^2}} \left(\frac{4 \times 10^3}{\epsilon}\right)^{\frac{1.28 \times 10^8 (n-1)}{\epsilon^2}} \\ &\geq e^{c\epsilon^{-2} n \log \frac{1}{\epsilon}} \end{aligned}$$

(where c is an absolute constant), then for any m-dimensional centrally symmetric convex body C, there is a subspace R^n of R^m such that $C \cap R^n$ is an ϵ-sphere. In other words, we have proven:

Theorem 7.2 (V.D. Milman [1]). *There is an absolute constant c such that Dvoretzky's theorem holds with*

$$M(n, \epsilon) \leq e^{c\epsilon^{-2} n \log \frac{1}{\epsilon}}.$$

§9. Some Historical Remarks

Dvoretzky's theorem is doubtlessly one of the deepest results in mathematics. Indeed, the extraordinary difficulty of Dvoretzky's original proof prompted V.D. Milman and G. Schechtman [1] to exclaim: "Dvoretzky's original proof was very complicated and understood only by a few people."

The first success in simplifying Dvoretzky's proof was obtained by V.D. Milman [1] in 1971. The main idea of Milman's proof was to exploit a certain property of *Haar measure* on high-dimensional homogeneous spaces, a property that is now called

The Concentration Phenomenon: *Given $\{\Re, \rho\}$ a compact metric space with a Borel probobility measure μ, define the concentration function $\alpha(\Re, \epsilon)$ for $\epsilon > 0$ by*

$$\alpha(\Re, \epsilon) = 1 - \inf\{\mu(A_\epsilon) : \ \mu(A) \geq \tfrac{1}{2}, \ A \subseteq \Re \text{ Borel set}\},$$

where

$$A_\epsilon = \{x \in \Re : \ \rho(x, A) \leq \epsilon\}.$$

Then, for some very natural families of spaces and measures, $\alpha(\Re, \epsilon)$ is extremely small.

For example, it follows from *Levy's isoperimetric inequality* that for $\Re = \partial(B)$, ρ the *geodesic metric* on $\partial(B)$, and μ the normalized rotationally invariant measure on $\partial(B)$,

$$\alpha(\partial(B), \epsilon) \leq \sqrt{\frac{\pi}{8}} e^{-\frac{\epsilon^2 n}{2}}.$$

Besides its usefullness in proving Dvoretzky's theorem, the concentration phenomenon plays a very important role in the further study of finite-dimensional normed spaces. Moreover, this phenomenon is connected with many other important mathematical problems; e.g., the problem of obtaining estimates on the kissing numbers of high-dimensional spheres.

Later, several other proofs of Dvoretzky's theorem were found: the functional analytic proofs of J.L. Krivine [1] and L. Tzafriri [1], the measure-theoretic proofs of T. Figiel [1] and A. Szankowski [1], and, most recently, the probabilistic proof of Y. Gordon [1]. However, these new proofs typically employ sophisticated techniques or rest on some other deep result, and so are not elementary and easy to understsand. The proof presented here is based upon the one in A. Szankowski [1].

The generalization of Dvoretzky's theorem from centrally symmetric convex bodies to arbitrary convex bodies was done by A. Dvoretzky himself. V.D. Milman's proof also handles Dvoretzky's theorem in this generality. In A. Dvoretzky [2] and D. Larman and P. Mani [1], one can find further generalizations concerned with projections rather than sections.

With regard to upper bounds on $M(n,\epsilon)$, A. Dvoretzky [1] first showed that

$$M(n,\epsilon) \leq e^{c\epsilon^{-2}n^2 \log^2 n}.$$

Later, V.D. Milman [1] proved that

$$M(n,\epsilon) \leq e^{c\epsilon^{-2}n \log \frac{1}{\epsilon}} \tag{7.23}$$

and, through examination of the special convex body W^m, noted that

$$M(n,\epsilon) \geq e^{cn \log \frac{1}{\epsilon}}. \tag{7.24}$$

A comparison of (7.23) and (7.24) shows that the exponential order of Milman's upper bound cannot be much improved.

Since $\{R^n, \|\cdot\|_C\}$ is a *Banach space* for every n-dimensional centrally symmetric convex body C, many generalizations and applications of Dvoretzky's theorem have been found in Functional Analysis. In fact, the so-called Local Theory of Banach Spaces is based on this theorem. For detailed information about this direction of research, we refer the reader to J. Lindenstrauss [1], V.D. Milman [3], J. Lindenstrauss and V.D. Milman [1], and G. Pisier [1].

Bibliography

Aleksandrov, A.D. [1]. Die Erweiterung zweier Lehrsätze Minkowskis über die konvexen Polyeder auf beliebige konvexe Flächen, *Mat. Sbornik N. S.* **3** (1938), 27–46. [**2**]. Almost everywhere existence of the second differential of a convex function and some properties of convex surfaces connected with it (Russian), *Uchenye Zapiski Leningrad. Gos. Univ. Math. Ser.* **6** (1939), 3–35. [**3**]. On tiling space by polytopes, *Vestnik Leningrad Univ. Ser. Mat. Fiz. Him.* **9** (1954), 33–43. [**4**]. *Konvexe Polyeder*, Akademie-Verlag, Berlin, 1958.

Baire, R. [**1**]. Sur les fonctions de variables réelles, *Ann. Mat. Pura Appl.* **3** (1899), 1–122.

Ball, K. [**1**]. Cube slicing in R^n, *Proc. Amer. Math. Soc.* **97** (1986), 465–473. [**2**]. Some Remarks on the Geometry of Convex Sets, Lecture Notes in Math. No. 1137, 224–231. Springer-Verlag Berlin, 1988.

Betke, U. and Gritzmann, P. [**1**]. Über L. Fejes Tóth's Wurstvermutung in kleinen Dimensionen, *Acta Math. Hungar.* **43** (1984), 299–307.

Betke, U., Gritzmann, P., and Wills, J.M. [**1**]. Slices of L. Fejes Tóth's sausage conjecture, *Mathematika* **29** (1982), 194–201.

Betke, U., Henk, M., and Wills, J.M. [**1**]. Finite and infinite packings, *J. Reine Angew. Math.* **453** (1994), 165–191.

Blaschke, W. [**1**]. *Kreis und Kugel* Veit & Co., Leipzig, 1916; Chelsea, New York; reprinted 1949; de Gruyter, Berlin, 1956.

Boltjanskii, V. and Gohberg, I. [**1**]. *Results and Problems in Combinatorical Geometry*, Cambridge University Press, Cambridge, 1985.

Bonnesen, T. and Fenchel, W. [**1**]. *Theorie der Konvexen Körper*, Springer-Verlag, Berlin, 1934, 1974; Chelsea, New York; reprinted 1948.

Böröczky Jr., K. [**1**]. About four-ball packings, *Mathematika* **40** (1993), 226–232. [**2**]. Mean projections and finite packings of convex bodies, *Monatsh. Math.* **118** (1994), 41–54.

Böröczky Jr., K. and Henk, M. [**1**]. Radii and the sausage conjecture, *Can. Math. Bull.* **38** (1995), 156–166.

Borsuk, K. [**1**]. Über die Zerlegung einer Euklidischen n-dimensionalen Vollkugel in n Mengen, *Verh. Internat. Math-Kongr. Zürich* (1932), Vol. **2**, p. 192. [**2**]. Drei Sätze über die n-dimensionalen Euklidische Sphäre, *Fund. Math.* **20** (1933), 177–190.

Bourgain, J. [**1**]. On the Busemann-Petty problem for perturbations of the ball, *Geom. Func. Analysis* **1** (1991), 1–13.

Busemann, H. [**1**]. Volume in terms of concurrent cross-sections, *Pacific J. Math.* **3** (1953), 1–12. [**2**]. *Convex Surfaces*, Interscience, New York, 1958. [**3**]. Volumes and areas of cross sections, *Amer. Math. Monthly* **67** (1960), 248–250; Correction **67** (1960), 671.

Busemann, H. and Feller, W. [**1**]. Krümmungseigenschaften konvexer Flächen, *Acta Math.* **66** (1936), 1–47.

Busemann, H. and Petty, C.M. [**1**]. Problem 1, Problems on convex bodies, *Math. Scand.* **4** (1965), 88–94.

Conway, J.H. and Sloane, N.J.A. [**1**]. *Sphere Packings, Lattices and Groups*, Springer-Verlag, New York, 1988.

Coxeter, H.S.M. [**1**]. An upper bound for the number of equal nonoverlapping spheres that can touch another of the same size, *Proc. Symp. Pure Math.* **7** (1963), 53–71.

Croft, H.T., Falconer, K.J., and Guy, R.K. [**1**]. *Unsolved Problems in Geometry*, Springer-Verlag, New York, 1991.

Danzer, L. [**1**]. Über Durchschnitteigenschaften n-dimensionalen Kugelfamilien, *J. Reine Angew. Math.* **209** (1960), 181–201.

Danzer, L., Laugwitz, D., and Lenz, H. [**1**]. Über das Löwnersche Ellipsoid und sein Analogon unter den einem Eikörper einbeschriebenen Ellipsoiden, *Arch. Math.* **8** (1957), 214–219.

Dekster, B.V. [1]. Borsuk's covering for blunt bodies, *Arch. Math.* (*Basel*) **51** (1988), 87–91.

Delone, B.N. [1]. Sur la partition réguliére de Léspace à 4 dimensions, *Izv. Akad. Nauk SSSR Otdel. Fiz.-mat. Nauk* **7** (1929), 79–110, 145–164.

Dvoretzky, A. [1]. Some results on convex bodies and Banach spaces, *Proc. Symp. on Linear Spaces* (1961), 123–160. [**2**]. Some near-sphericity results, *Proc. Symp. Pure Math.* **7** (1963), 203–210.

Dvoretzky, A. and Rogers, C.A. [1]. Absolute and unconditional convergence in normed linear spaces, *Proc. Nat. Acad. Sci. USA* **36** (1950), 192–197.

Eggleston, H.G. [1]. Covering a three-dimensional set with sets of smaller diameter, *J. London Math. Soc.* **30** (1955), 11–24. [**2**]. *Convexity*, Cambridge University Press, Cambridge, 1958.

Engel, P. [1]. On the symmetry classification of the four-dimensional parallelohedra, *Z. Kristallographie* **200** (1992), 199–213. [**2**]. Geometric crystallography, *Handbook of Convex Geometry* (P.M. Gruber and J.M. Wills, eds.), 989–1041. North-Holland, Amsterdam, 1993.

Erdös, P., Gruber, P.M., and Hammer, J. [1]. *Lattice Points*, Longman, Essex, 1989.

Fáry, I. and Makai, E. [1]. Research problems, *Period. Math. Hungar.* **14** (1983), 111–114. [**2**]. Problem, *Intuitive Geometry* (K. Böröczky and G. Fejes Tóth, eds.), 694–695. North-Holland, Amsterdam, 1987.

Fedorov, E.S. [1]. An introduction to the theory of figures (in Russian), *Zapiski. Imper. S.-Petersburgskago Miner. Obsćestva* **21** (1885), 1–278.

Fejes Tóth, G., Gritzmann, P., and Wills, J.M. [1]. Finite sphere packing and sphere covering, *Discrete Comp. Geom.* **4** (1989), 19–40.

Fejes Tóth, G. and Kuperberg, W. [1]. Packing and covering with convex sets, *Handbook of Convex Geometry* (P.M. Gruber and J.M. Wills, eds.), 799–860. North-Holland, Amsterdam, 1993.

Fejes Tóth, L. [1]. Über die dichteste Kreislagerung und dünnste Kreisüberdeckung, *Comment. Math. Helv.* **23** (1949), 342–349. [**2**]. Some packing and covering theorems, *Acta Sci. Math. Szeged.* **12** (1950), 62–67. [**3**]. *Lagerungen in der Ebene, auf der Kugel und im Raum*, Springer-Verlag, Berlin, 1972. [**4**]. Research problem 13, *Period. Math. Hungar.* **6** (1975), 197–199.

Fenchel, W. and Jessen, B. [1]. Mengenfunktion und konvexe Körper, *Danske Vid. Selsk. Mat.-Fys. Medd.* **16** (1938), 1–31.

Figiel, T. [1]. A short proof of Dvoretsky's theorem on almost spherical sections, *Compositio Math.* **33** (1976), 297–301.

Figiel, T., Lindenstrauss, J., and Milman, V.D. [1]. The dimension of almost spherical sections of convex bodies, *Acta Math.* **139** (1977), 53–94.

Frankl, P. and Wilson, R.M. [1]. Intersection theorems with geometric consequences, *Combinatorica* **1** (1981), 357–368.

Gale, D. [1]. On inscribing n-dimensional sets in a regular n-simplex, *Proc. Amer. Math. Soc.* **4** (1953), 222–225.

Gandini, P.M. and Wills, J.M. [1]. On finite sphere packings, *Math. Pannonica* **3** (1992), 19–29.

Gandini, P.M. and Zucco, A. [1]. On the sausage catastrophe in 4-space, *Mathematika* **39** (1992), 274–278.

Gardner, R.J. [1]. Intersection bodies and the Busemann-Petty problem, *Trans. Amer. Math. Soc.* **342** (1994), 435–445. [2]. A positive answer to the Busemann-Petty problem in three dimensions, *Ann. Math.* **140** (1994), 435–447.

Gauss, C.F. [1]. Untersuchungen über die Eigenschaften der positiven ternären quadratischen Formen von Ludwig August Seeber, *J. Reine Angew. Math.* **20** (1840), 312–320.

Giannopoulos, A.A. [1]. A note on a problem of H. Busemann and C.M. Petty concerning sections of convex bodies, *Mathematika* **37** (1990), 239–244.

Gietz, M. [1]. A note on a problem of Busemann, *Math. Scand.* **25** (1969), 145–148.

Gordon, Y. [1]. Some inequalities for Gaussian processes and applications, *Israel J. Math.* **50** (1985), 265–289.

Grinberg, E.L. and Riven, I. [1]. Infinitesimal aspects of the Busemann-Petty problem, *Bull. London Math. Soc.* **22** (1990), 478–488.

Gritzmann, P. [1]. Finite packing of equal balls, *J. London Math. Soc.* **33** (1986), 543–553.

Gritzmann, P. and Wills, J.M. [1]. Finite packing and covering, *Handbook of Convex Geometry* (P.M. Gruber and J.M. Wills, eds.), 861–897. North-Holland, Amsterdam, 1993.

Groemer, H. [**1**]. Über die Einlagerungen von Kreisen in einem konvexen Bereich, *Math. Z.* **73** (1960), 285–294. [**2**]. Abschätzungen für die Anzahl der konvexen Körper, die einen konvexen Körper berühren, *Monatsh. Math.* **65** (1961), 74–81. [**3**]. Über die dichteste gitterförmige Lagerung kongruenter Tetraeder, *Monatsh. Math.* **66** (1962), 12–15. [**4**]. Über Zerlegungen des Euklidischen Raumes, *Math. Z.* **79** (1962), 364–375.

Grothendieck, A. [**1**]. Sur certaines classes de s̀uites dans les espaces de Banach et le théorèm de Dvoretsky-Rogers, *Bol. Soc. Mat. San Paulo* **8** (1953), 83–110.

Gruber, P.M. [**1**]. Die meisten konvexen Körper sind glatt, aber nicht zu glatt, *Math. Ann.* **229** (1977), 259–266. [**2**]. Typical convex bodies have surprisingly few neighbours in the densest lattice packings, *Studia Sci. Math. Hungar.* **21** (1986), 163–173. [**3**]. Minimal ellipsoids and their duals, *Rend. Circ. Mat. Palermo*, (2) **37** (1988), 35–64. [**4**]. Baire categories in convexity, *Handbook of Convex Geometry* (P.M. Gruber and J.M. Wills, eds.), 1327–1346. North-Holland, Amsterdam, 1993.

Gruber, P.M. and Höbinger, J. [**1**]. Kennzeichnungen von Ellipsoiden mit Anwendungen, *Jahrbuch Uberblicke Math.* (1976), 9–29.

Gruber, P.M. and Lekkerkerker, C.G. [**1**]. *Geometry of Numbers* (2nd ed.), North-Holland, Amsterdam, 1987.

Grünbaum, B. [**1**]. A simple proof of Borsuk's conjecture in three dimensions, *Proc. Cambridge Philos. Soc.* **53** (1957), 776–778. [**2**]. On a conjecture of H. Hadwiger, *Pacific J. Math.* **11** (1961), 215–219. [**3**]. Borsuk's problem and related questions, *Proc. Symp. Pure Math.* **7** (1963), 271–284.

Grünbaum, B. and Shephard, G.C. [**1**]. Tiling with congruent tiles, *Bull. Amer. Math. Soc.* **3** (1980), 951–973. [**2**]. *Tilings and Patterns*, Freeman, New York, 1986.

Hadwiger, H. [**1**]. Überdeckung einer Menge durch Mengen kleineren Durchmesse, *Comment. Math. Helv.* **18** (1945/46), 73–75; **19** (1946/47), 72–73. [**2**]. Einfache Herleitung der isoperimetrischen Ungleichung für abgeschlossene Punktmengen, *Math. Ann.* **124** (1952), 158–160. [**3**]. Über Treffanzahlen bei translationsgleichen Eikörper, *Arch. Math.* **8** (1957), 211–213. [**4**]. *Vorlesungen über Inhalt, Oberfläche und Isoperimetrie*, Springer-Verlag, Berlin, 1957.

Hajós, G. [**1**]. Über einfache und mehrfache Bedeckung des n-dimensionalen Raumes mit einem Würfelgitter, *Math. Z.* **47** (1942), 427–467.

Hales, T.C. [**1**]. The status of the Kepler conjecture, *Math. Intelligencer* **16**(3) (1994), 47–58.

Halmos, P. [**1**]. *Measure Theory*, Van Nostrand, Princeton, NJ, 1950.

Hamaker, W. [**1**]. Factoring groups and tiling space, *Aequationes Math.* **9** (1973), 145–149.

Hamaker, W. and Stein, S.K. [**1**]. Splitting groups by integers, *Proc. Amer. Math. Soc.* **46** (1974), 322–324.

Henk, M. [**1**]. *Finite and Infinite Packings*, Habilitationsschrift, Univ. Siegen, 1995.

Heppes, A. [**1**]. On the partitioning of three-dimensional point sets into sets of smaller diameter (in Hungarian), *Magyar Tud. Akad. Mat. Fiz. Oszt. Kösl.* **7** (1957), 413–416.

Heppes, A. and Révész, P. [**1**]. Zum Borsukschen Zerteilungsproblem, *Acta Math. Sci. Hungar.* **7** (1956), 159–162.

Hilbert, D. [**1**]. Mathematische Probleme, *Arch. Math. Phys.* **1** (1901), 44–63.

Hlawka, E. [**1**]. Ausfüllung und Überdeckung konvexer Körper durch konvexe Körper, *Monatsh. Math.* **53** (1949), 81–131.

Holmes, R. [**1**]. *Geometric Functional Analysis and its Application*, Springer-Verlag, Berlin, 1975.

Hoppe, R. [**1**]. Bemerkungen der Redaktion (von R. Hoppe) zu C. Bender, Bestimmung der grössten Anzahl gleich grosser Kugeln, welche sich auf eine Kugel von demselben Radius wie die übrigen auflegen lassen, *Grunert Arch. Math. Phys.* **56** (1874), 302–306.

Hoylman, D.J. [**1**]. The densest lattice packing of tetrahedra, *Bull. Amer. Math. Soc.* **76** (1970), 135–137.

Hsiang, W.Y. [**1**]. On the sphere packing problem and the proof of Kepler's conjecture, *Internat. J. Math.* **4** (1993), 739–831. [**2**]. A rejoinder to Hales's article, *Math. Intelligencer* **17** (1) (1995), 35–42.

Kabatjanski, G.A. and Levenštein, V.I. [**1**]. Bounds for packings on a sphere and in space, *Problemy Peredachi Informatsii* **14** (1978), 3–25.

Kahn, J. and Kalai, G. [**1**]. A counterexample to Borsuk's conjecture, *Bull. Amer. Math. Soc.* **29** (1993), 60–62.

Kepler, J. [1]. *Strena seu de nive sexangula*, Tampach, Frankfurt, 1611; *Gesammelte Werke* **4**, 259–80. Beck, München 1941; Clarendon Press, Oxford, 1966.

Klee, V. [1]. Some new results on smoothness and rotundity in normed linear spaces, *Math. Ann.* **139** (1959), 51–63.

Kleinschmidt, P., Pachner, U., and Wills, J.M. [1]. On L. Fejes Tóth's sausage conjecture, *Israel J. Math.* **47** (1984), 216–226.

Klima, V. and Netuka, I. [1]. Smoothness of a typical convex function, *Czechoslovak Math. J.* **31** (1981), 569–572.

Krivine, J.L. [1]. Sous-espaces de dimension finie des espaces de Banach réticulés, *Ann. Math.* **104** (1976), 1–29.

Larman, D.G. [1]. Open problem 6, *Convexity and Graph Theory* (M. Rosenfeld and J. Zaks, eds), *Ann. Discrete Math.* **20** (1984), 336.

Larman, D.G. and Mani, P. [1]. Almost ellipsoidal sections and projections of convex bodies, *Proc. Cambridge Philos. Soc.* **77** (1975), 529–546.

Larman, D.G. and Rogers, C.A. [1]. The existence of a centrally symmetric convex body with central sections that are unexpectedly small, *Mathematika* **22** (1975), 164–175.

Lassak, M. [1]. An estimate concerning Borsuk partition problem, *Bull. Acad. Polon. Sci. Sér. Sci. Math.* **30** (1982), 449–451.

Leech, J. [1]. The problem of the thirteen spheres, *Math. Gaz.* **40** (1956), 22–23.

Leichtweiß, K. [1]. Über die affine Exzentrizität konvexer Körper, *Arch. Math.* **10** (1959), 187–199. [2]. *Konvexe Mengen*, Deuts. Verl. Wiss., Berlin, 1980; Springer-Verlag, Berlin, 1980.

Lenz, H. [1]. Zur Zerlegung von Punktmengen in solche kleineren Durchmessers, *Arch. Math.* **6** (1955), 413–416. [2]. Zerlegung ebener Bereiche in konvex Zellen von möglichst kleinen Durchmesser, *Jber. Deutsch. Math. Verein* **58** (1956), 87–97. [3]. Über die Bedeckung ebener Punktmengen durch solche kleineren Durchmessers, *Arch. Math.* **7** (1956), 34–40.

Levenštein, V.I. [1]. On bounds for packings in n-dimensional Euclidean space, *Dokl. Akad. Nauk SSSR* **245** (1979), 1299–1303.

Lindenstrauss, J. [1]. Almost spherical sections, their existence and their applications, *Jber. Deutsch. Math. Verein., Jubiläumstagung 1990*, 39–61, Teubner, Stuttgart, 1992.

Lindenstrauss, J. and Milman, V.D. [**1**]. The local theory of normed spaces and its applications to convexity, *Handbook of Convex Geometry* (P.M. Gruber and J.M. Wills, eds.), 1149–1220. North-Holland, Amsterdam, 1993.

Lindenstrauss, J. and Tzafriri, L. [**1**]. *Classical Banach Spaces*, Lecture Notes in Math. No. 338, Springer-Verlag, Berlin, 1973.

Lutwak, E. [**1**]. Intersection bodies and dual mixed volumes, *Adv. Math.* **71** (1988), 232–261. [**2**]. Selected affine isoperimetric inequalities, *Handbook of Convex Geometry* (P.M. Gruber and J.M. Wills, eds.), 151–176. North-Holland, Amsterdam, 1993.

McMullen, P. [**1**]. Space tiling zonotopes, *Mathematika* **22** (1975), 202–211. [**2**]. Convex bodies which tile space by translation, *Mathematika* **27** (1980), 113–121.

Milman, V.D. [**1**]. A new proof of the theorem of A. Dvoretzky on sections of convex bodies, *Func. Anal. Appl.* **5** (1971), 28–37 (translated from Russian). [**2**]. The concentration phenomenon and linear structure of finite dimensional normed spaces, *Proc. Internat. Congr. Math. Berkeley* (1986), 961–975. [**3**]. Dvoretzky theorem-thirty years later, *Geom. Func. Analysis* **2** (1992), 455–479.

Milman, V.D. and Schechtman, G. [**1**]. *Asymptotic Theory of Finite Dimensional Normed Spaces*, Lecture Notes in Math. No. 1200, Springer-Verlag, Berlin, 1986.

Minkowski, H. [**1**]. *Geometrie der Zahlen*, Teubner, Leipzig, 1896; reprinted Chelsea, New York 1953. [**2**]. Allgemeine Lehrsätze über konvexen Polyeder, *Nachr. K. Ges. Wiss. Göttingen, Math.-Phys. KL* (1897), 198–219. [**3**]. Volumen und Oberfläche, *Math. Ann.* **57** (1903), 447–495. [**4**]. Dichteste gitterförmige Lagerung kongruenter Körper, *Nachr. K. Ges. Wiss. Göttingen, Math.-Phys. KL* (1904), 311–355.

Müller, C. [**1**]. *Spherical Harmonics*, Lecture Notes in Math. No. 17, Springer-Verlag, Berlin, 1966.

Odlyzko, A.M. and Sloane, N.J.A. [**1**]. New bounds on the number of unit spheres that can touch a unit sphere in n dimensions, *J. Comb. Theory* (*A*) **26** (1979), 210–214.

Osgood, W.F. [**1**]. Zweite Note über analytische Funktionen mehrer Veränderlichen, *Math. Ann.* **53** (1900), 461–463.

Oxtoby, J.C. [**1**]. *Measure and Category*, Springer-Verlag, Berlin, 1971.

Pál, J. [**1**]. Über ein elementares Variationsproblem, *Danske Vid. Selsk. Math.-Fys. Medd.* **3**(2) (1920).

Papadimitrakis, M. [1]. On the Busemann-Petty problem about convex centrally symmetric bodies in R^n, *Mathematika* **39** (1992), 258–266.

Perkal, J. [1]. Sur la subdivision des ensembles en parties da diamètre inférieur, *Colloq. Math.* **1** (1947), 45.

Petty, C.M. [1]. Projection bodies, *Proc. Coll. Convexity* (1967), 234–241.

Pisier, G. [1]. *The Volume of Convex Bodies and Banach Space Geometry,* Cambridge University Press, Cambridge, 1989.

Rankin, R.A. [1]. The closest packing of spherical caps in n dimensions, *Proc. Glasgow Math. Assoc.* **2** (1955), 139–144.

Reidemeister, K. [1]. Über die singulären Randpunkte eines konvexen Körpers, *Math. Ann.* **83** (1921), 116–118.

Riesling, A.S. [1]. Borsuk's problem in three-dimensional spaces of constant curvature, *Ukr. Geom. Sbornik* **11** (1971), 78–83.

Rogers, C.A. [1]. The closest packing of convex 2-dimensional domains, *Acta Math.* **86** (1951), 309–321. [**2**]. Symmetrical sets of constant width and their partitions, *Mathematika* **18** (1971), 105–111. [**3**]. *Packing and Covering,* Cambridge University Press, Cambridge, 1964.

Rush, J.A. [1]. A lower bound on packing density, *Invent. Math.* **98** (1989), 499–509.

Schneider, R. [1]. Zu einem Problem von Shephard über die Projektionen konvexer Körper, *Math. Z.* **101** (1967), 71–82. [**2**]. On the curvatures of convex bodies, *Math. Ann.* **240** (1979), 177–181. [**3**]. *Convex Geometry: The Brunn-Minkowski Theory,* Cambridge University Press, Cambridge, 1993.

Schramm, O. [1]. Illuminating sets of constant width, *Mathematika* **35** (1988), 180–189.

Schulte, E. [1]. Tilings, *Handbook of Convex Geometry* (P.M. Gruber and J.M. Wills, eds.), 899–932. North-Holland, Amsterdam, 1993.

Schütte, K. and van der Waerden, B.L. [1]. Das Problem der dreizehn Kugeln, *Math. Ann.* **125** (1953), 325–334.

Shephard, G.C. [1]. Shadow systems of convex bodies, *Israel J. Math.* **2** (1964), 229–236. [**2**]. Approximation problems for convex polyhedra, *Mathematika* **11** (1964), 9–18. [**3**]. Space tiling zonotopes, *Mathematika* **21** (1974), 261–269.

Sierpiński, W. [1]. *Elementary Theory of Numbers,* Warschau, 1964.

Stein, S.K. [**1**]. Factoring by subsets, *Pacific J. Math.* **22** (1967), 523–541. [**2**]. A symmetric star body that tiles but not as a lattice, *Proc. Amer. Math. Soc.* **36** (1972), 543–548. [**3**]. Tiling, packing and covering by clusters, *Rocky Mountain J. Math.* **16** (1986), 277–321.

Steiner, J. [**1**]. Einfache Beweise der isoperimetrischen Hauptsätze, *J. Reine Angew. Math.* **18** (1838), 281–296.

Štogrin, M.I. [**1**]. Regular Dirichlet-Voronoi partitions for the second triclinic group, *Proc. Steklov Inst. Math.* **123** (1973).

Swinnerton-Dyer, H.P.F. [**1**]. Extremal lattices of convex bodies, *Proc. Cambridge Philos. Soc.* **49** (1953), 161–162.

Szabó, S. [**1**]. On mosaics consisting of multidimensional crosses, *Acta Math. Acad. Sci. Hungar.* **39** (1981), 191–203. [**2**]. Rational tilings by n-dimensional crosses, *Proc. Amer. Math. Soc.* **87** (1983), 212–222. [**3**]. A star polyhedron that tiles but not as a fundamental region, *Colloq. Math. Soc. János Bolyai* **48** (1985), 531–544.

Szankowski, A. [**1**]. On Dvoretzky's theorem on almost spherical sections of convex bodies, *Israel J. Math.* **17** (1974), 325–338.

Tzafriri, L. [**1**]. On Banach spaces with unconditional Bases, *Israel J. Math.* **17** (1974), 84–93.

Venkov, B.A. [**1**]. On a class of Euclidean polytopes, *Vestnik Leningrad Univ. Ser. Mat. Fiz. Him.* **9** (1954), 11–31.

Watson, G.L. [**1**]. The number of minimum points of a positive quadratice form, *Dissertationes Math.* **84** (1971), 1–42.

Wegner, G. [**1**]. Über endliche Kreispackungen in der Ebene, *Studia Sci. Math. Hungar.* **21** (1986), 1–28.

Wills, J.M. [**1**]. Research problem 30, *Period. Math. Hungar.* **13** (1982), 75–76. [**2**]. Research Problem 35, *Period. Math. Hungar.* **14** (1983), 312–314. [**3**]. On the density of finite packings, *Acta Math. Hungar.* **46** (1985), 205–210. [**4**]. On large lattice packings of spheres, *Geom. Dedicata*, to appear.

von Wolff, M.R. [**1**]. A star domain with densest admissible point set not a lattice, *Acta Math.* **108** (1962), 53–60.

Wyner, J.M. [**1**]. Capabilities of bounded discrepancy decoding, *Bell System Tech. J.* **44** (1965), 1061–1122.

Zamfirescu, T. [**1**]. Nonexistence of curvature in most points of most convex bodies, *Math. Ann.* **252** (1980), 217–219. [**2**]. The curvature of most convex surfaces vanishes almost everywhere, *Math. Z.* **174** (1980), 135–139. [**3**]. Many endpoints and few interior points of geodesics, *Invent. Math.* **69** (1982), 253–257. [**4**]. Using Baire categories in geometry, *Rend. Sem. Mat. Univ. Politecn. Torino* **43** (1985), 67–88.

Zaremba, S.K. [**1**]. Covering problems concerning Abelian groups, *J. London Math. Soc.* **27** (1952), 242–246.

Zhang, G. [**1**]. Intersection bodies and the Busemann-Petty problem in R^4, *Ann. Math.* **140** (1994), 331–346. [**2**]. Centered bodies and dual mixed volumes, *Trans. Amer. Math. Soc.* **345** (1994), 777–801.

Zong, C. [**1**]. *Packing and Covering*, Ph.D thesis, Technische Universität Wien, 1993. [**2**]. An example concerning the translative kissing number of a convex body, *Discrete Comput. Geom.* **12** (1994), 183–188. [**3**]. On a conjecture of Croft, Falconer and Guy on finite packings, *Arch. Math.* **64** (1995), 269–272. [**4**]. Some remarks concerning kissing numbers, blocking numbers and covering numbers, *Period. Math. Hungar.* **30** (1995), 233–238. [**5**]. The kissing numbers of tetrahedra, *Discrete Comput. Geom.* **15** (1996), 251–264.

Index

Universitext *(continued)*

Rubel/Colliander: Entire and Meromorphic Functions
Sagan: Space-Filling Curves
Samelson: Notes on Lie Algebras
Schiff: Normal Families
Shapiro: Composition Operators and Classical Function Theory
Simonnet: Measures and Probability
Smith: Power Series From a Computational Point of View
Smoryński: Self-Reference and Modal Logic
Stillwell: Geometry of Surfaces
Stroock: An Introduction to the Theory of Large Deviations
Sunder: An Invitation to von Neumann Algebras
Tondeur: Foliations on Riemannian Manifolds
Zong: Strange Phenomena in Convex and Discrete Geometry